SCIENCE AND LIBERTY

PATIENT CONFIDENCE IN THE ULTIMATE JUSTICE OF THE PEOPLE

by

John L. Cordani Jr.

Series in Politics
VERNON PRESS

Copyright © 2022 Vernon Press, an imprint of Vernon Art and Science Inc, on behalf of the author.

All rights reserved. No part of this publication may be reproduced, stored in a retrieval system, or transmitted in any form or by any means, electronic, mechanical, photocopying, recording, or otherwise, without the prior permission of Vernon Art and Science Inc.
www.vernonpress.com

In the Americas:	*In the rest of the world:*
Vernon Press	Vernon Press
1000 N West Street, Suite 1200,	C/Sancti Espiritu 17,
Wilmington, Delaware 19801	Malaga, 29006
United States	Spain

Series in Politics

Library of Congress Control Number: 2021941141

ISBN: 978-1-64889-325-4

Also available: 978-1-64889-279-0 [Hardback]; 978-1-64889-312-4 [PDF, E-Book]

Product and company names mentioned in this work are the trademarks of their respective owners. While every care has been taken in preparing this work, neither the authors nor Vernon Art and Science Inc. may be held responsible for any loss or damage caused or alleged to be caused directly or indirectly by the information contained in it.

Every effort has been made to trace all copyright holders, but if any have been inadvertently overlooked the publisher will be pleased to include any necessary credits in any subsequent reprint or edition.

Cover design by Vernon Press.

Painting on the left: *The Alchymist, In Search of the Philosopher's Stone, Discovers Phosphorus, and prays for the successful Conclusion of his operation, as was the custom of the Ancient Chymical Astrologers*, by Joseph Wright of Derby (1734–1797), 1771.
Painting on the right: *Thomas Jefferson (right), Benjamin Franklin (left), and John Adams (center) meet at Jefferson's lodgings, on the corner of Seventh and High (Market) streets in Philadelphia, to review a draft of the Declaration of Independence. 1 photomechanical print: halftone, color (postcard made from painting). Postcard published by The Foundation Press, Inc., 1932. Reproduction of oil painting from the artist's series: The Pageant of a Nation.* Artist: Jean Leon Gerome Ferris (1863–1930).
Background: https://pixabay.com/photos/paper-document-old-writing-vintage-3212015/

Table of contents

1. **Introduction**   1
2. **Founding Principles**   13
3. **The Rule of Scientists**   33
4. **The Excellent Servant and the Terrible Master**   63
5. **The False Idols**   85
6. **Ideology Fills the Vacuum**   123
7. **Conclusion: Science and Liberty**   147

Bibliography   157

Index   165

1.
Introduction

My first job in high school was working at a movie theater. It was a small, independent theater with a handful of screens. Employees like me did a little bit of everything, as needed. I popped the popcorn and slung concessions. I sold tickets and ripped tickets. I cleaned the theaters and did the mid-show walkthroughs using flashlights with plastic cones on them. I loved movies then and have ever since.

As a small theater, sometimes we did not get to run the big blockbusters. By and large, the big chain theater on the other side of town got them. But there was one year where our theater was not only getting to run the new, big James Bond flick; we were also getting to host an early screening about a week before the official release date. While I did not know all the ins and outs, a lot of the special tickets to the screening were part of promotions around town. You could get one by winning a radio contest. Since Bond drove a BMW in the movie, I think the local dealership got a few promotional tickets for its customers.

I asked my manager for a shift to work that night. As the one and only evening showtime approached, people lined up outside to get good seats, excited for the opportunity. Before it was time to let everyone in, the manager went outside and returned with someone from the line. He was a film critic from our local newspaper. I served the guy some concessions that the manager comp'd, as the two of them talked about various "films." When the critic was all set and he was ushered into the theater to have his choice of seat, the manager turned to me and said, "ok, now go let the *hoi polloi* in." I opened the doors and remember the people having a very nice time that night. I even caught a few of the action set pieces standing at the back door of the theater with my cone-equipped flashlight.

As a kid, I did not know what "*hoi polloi*" meant. That might have been the first time I heard the term, or at least it was the first time it piqued my interest. I understood what the manager wanted from the context, but I looked it up later. *Hoi polloi* is an idiom borrowed from Greek. Today, it is generally used in a somewhat derogatory fashion to mean the masses, the common folk, the rabble. The theater manager and the movie critic writing for the local paper were not the *hoi polloi*. They were movie buffs. Better yet, they were film buffs. They could tell you what *films* were actually good and the deep layers of meaning created by *auteur*-directors. Looking back, the fact that the early screening that night was a *James Bond* movie makes the whole incident a bit odd. The movie critic was not like a food critic coming to a restaurant. He was not coming to criticize or even review our theater. He was just there to watch the movie and review it. The manager did not need to treat him special to get a nice review. More to the point, the movie he was there to see was not some arthouse film. It was a blockbuster. Like life, it was made for everyone to enjoy and have a say in. No *B.A.* in film studies required.

I was reminded of that event again in college. While I took my major in science, I loved learning about the ancient Greeks. I was intrigued by the concept of the city-state, citizenship, the wars fought when western civilization was in its infancy, the famous 300 Spartans fighting at Thermopylae, and the like. There I learned that the idea of a "*hoi polloi*" went back to Athens' great leader and general, Pericles. The ancient Greek historian, Thucydides, recorded Pericles famous Funeral Oration for those Athenian citizens who had died in the Peloponnesian War against Sparta. In the speech, Pericles contrasted the *polloi* (the People) with the *oligoi* (the few, as in an "oligopoly"). For Pericles, the *hoi polloi* was not derogatory. It was good; it was of-the-essence for a democracy like Athens. He said, "Our form of government is called a democracy because its administration is in the hands, not of the few [*hoi oligoi*], but of the whole people [*hoi polloi*]....Everyone is equal before the law."[1]

This belief is fundamental, and no one put it better than Pericles. In a few words, he had gotten to the heart of the matter. He understood the link between democracy, how a government is administered, and the ideal of equality in the eyes of the law. Of course, not all Athenians agreed with Pericles. Plato had watched his teacher and mentor, Socrates, unjustly executed at the hands of an Athenian jury. He believed that democracy was doomed to end in the *hoi polloi* in the worst sense—with the anarchy and tyranny of an inflamed, emboldened, and selfish rabble. Plato's ideal state was one run by a philosopher king, an expert who would know best what to do. On the other hand, Plato's student, Aristotle, turned out to be a bit more bullish on democracy like Pericles. Aristotle theorized that, if we could properly channel the people's power in a democracy, their collective judgment could be even better than "those who have special knowledge."[2]

While America's Founding Fathers would have agreed with Pericles and Aristotle, with some qualifications, today we seem to be siding with Plato more

---

[1] Thucydides, "Pericles Funeral Oration for Athenian War Dead," Rjgeib.com, Accessed May 28, 2021, https://www.rjgeib.com/thoughts/athens/athens.html
[2] Aristotle and B. Jowett, trans., *Aristotle's politics and poetics* (New York: Viking Press, 1974), Politics III. 11.

and more. Except now, the proposed kings are not philosophers; they are an idealized form of benevolent scientist.

Scientific Success

Since the Age of Enlightenment and the Industrial Revolution it helped create, the successes of science have been unparalleled. No one could dispute that, in the last two hundred years and with the help of science, we have learned more about the natural world and how to harness it to improve our lives than in all the other years of human existence combined. Harvard psychologist Steven Pinker ably marshalled, cataloged, and touted all of these undeniable benefits of science in his 2018 bestselling book, *Enlightenment Now: The Case for Reason, Science, Humanism and Progress*. His 576-page treatment makes good on its dust-jacket promise of "seventy-five jaw-dropping graphs, [by which] Pinker shows that life, health, prosperity, safety, peace, knowledge, and happiness are on the rise, not just in the West, but worldwide…[based on] the conviction that reason and science can enhance human flourishing."[3] Science has been a true success story with all the objective, chart-based data you could shake a stick at to show for it.

It is easy to see, then, how tempting it is for us to ask more of the scientist. On the one hand, channeled through business and academia, scientists have proven themselves capable of true progress by objective metrics. On the other hand, we have the seemingly eternal rancor of politicians who never seem to make any progress. The democratic pendulum seems to swing left or right every few years, but the same underlying disputes remain. The positions of both sides seem incommensurable and incapable of compromise, and sometimes it seems like we are stuck in some episode of the Twilight Zone. The People and, accordingly, their representatives just seem to fundamentally disagree on what is good, and right, and just for government and political administration. *Ah*, but if scientists have been able to use objective measurements to achieve concrete, undeniable *results*, why not give them a shot at making the calls in the political sphere? Sure, we can keep the politicians around and preserve our basic Constitutional institutions (Congress, the President, the courts), but maybe that all can just be a bit of play acting—smoke and mirrors for the scientist standing behind the curtain pulling the levers of administration, getting things done efficiently and in everyone's best interest. In short, let's just have our politicians "follow the science."

This idea—that science is the best mechanism for society to determine their normative or political values—is known as scientism. Even setting Plato aside,

---

[3] S. Pinker, *Enlightenment Now* (London: Penguin Books, 2019).

this concept is not a new one. It has been prevalent for quite some time in continental Europe and was imported to American academia around the turn of the Twentieth Century. And though it was once a fringe idea at least in the American system, scientism is gaining more and more ground today. For example, during one of the 2020 Presidential debates, then-candidate Joe Biden was asked whether he would lock down the United States economy if scientists concluded that it was necessary to contain the spread of coronavirus. He said, "I would shut it down. I would listen to the scientists."[4] During a later town-hall event, he reiterated, "I said I'd follow the science."[5] "We are following the science" became a mantra in the world's greatest democracies on how politicians responded to conditions of pandemic, sometimes despite protests from the scientists about the distinction between facts and values.[6]

Maybe desperate times called for desperate measures. Some may argue that a little sacrifice of democratic ideals could be tolerated to overcome an unprecedented crisis, and it is beyond the purpose of this book to speak to this recent crisis specifically. Nonetheless, the emergency responses recently deployed by democratic countries is just the visible tip of a scientism iceberg that has been forming for quite some time. It is part of a trend where scientists are not just called upon to opine on the facts, but to take responsibility for setting rules and policy based on those facts. More to the point, it is just one more instantiation of a world that is being silently absorbed into administrative law and the so-called administrative state.

Today, America's democracy is governed by a legion of public servants tasked with following the science every single day. No citizen ever cast a vote for them. And often, even the President cannot fire them unless they have been seriously derelict in performing their duties. For the most part, they do not appear on television or give interviews in the press to tell the People what they are doing and why. We have expert economists working at the Federal Reserve Bank and the Federal Trade Commission. Expert biologists, botanists, and veterinarians work for the United States Department of Agriculture. Expert physicists and chemists are at the United States Department of Energy; expert teachers are at

---

[4] CBS News, "Biden says he would shut the U.S. down if recommended by scientists," Cbsnews.com, Accessed: May 28, 2021, https://www.cbsnews.com/news/biden-shut-us-down-coronavirus-if-recommended-scientists/

[5] ABC News, "Read the full transcript of Joe Biden's ABC News town hall," ABCnews.com, October 15, 2020, Accessed: May 28, 2021, https://abcnews.go.com/Politics/read-full-transcript-joe-bidens-abc-news-town/story?id=73643517.

[6] A. Stevens, "Governments cannot just 'follow the science' on COVID-19," *Nature Human Behaviour* 4, 6 (2020): p.560.

the United States Department of Education; expert sociologists are at the United States Department of Housing and Urban Development. There is no need to continue; we all know how the list goes on. These scientists are all hard at work producing laws that they call regulations. In 1950, the federal government alone produced over 9,000 pages of them for the Code of Federal Regulations.[7] If you think that's a lot, then in 2019, the figure was 185,000 pages. Federal agencies have produced over 100,000 pages of regulations each and every year since 1980. A lot of these regulations carry criminal penalties, punishable by steep fines or imprisonment.

By and large, all of these scientists, bureaucrats, and other public servants are doing their level best to act in what they see as the best interests of the American people to produce good, worthwhile, and scientifically founded regulations. I am certainly not saying that there are no good ideas in those hundreds of thousands of pages of regulations. Nor am I saying that science does not have a role to play in policy, politics, or government. Before becoming a lawyer, I studied science as an undergraduate and worked in industry as a scientist. I love science. I think the scientific method was the single best process ever devised by humans to study the natural world. Having said that, however, the question addressed here is a different one. No matter how much I admire science, the issue is whether we are asking too much of science, for things that it cannot or should not be tasked with providing us. Is there is an inconsistency between what we have asked from science and what America's Constitution asked from the People themselves? And if so, is the Constitution's way is actually better when it comes to the realm of values, judgments, and the sphere of the political?

The answer seems to me to be that the unrestrained expansion of the sphere of science threatens both science and liberty. Suspicions of bias and tyranny are bred and nurtured when science is extended beyond its legitimate sphere, as a method for the determination of fact, to being an arbiter of human values and policy. Michel Gove, a UK politician who was one of the staunchest promoters of Brexit, was questioned ahead of that 2016 referendum about whether any economists supported Britain's exit from the European Union. He responded simply that the "people in this country have had enough of experts."[8] Gove turned out to be right. He had put his finger on an increasingly

---

[7] GW Regulatory Studies Center, "Reg Stats," Regulatorystudies.columbian.gwu.edu, Accessed: May, 28, 2021, https://regulatorystudies.columbian.gwu.edu/reg-stats
[8] H. Mance, "Britain has had enough of experts, says Gove," Ft.com, June 3, 2016, Accessed: May 28, 2021, https://www.ft.com/content/3be49734-29cb-11e6-83e4-abc22d5d108c.

prevalent populist movement that thinks "the science" cannot provide the sole justification for bureaucratic action.

A 2019 study by the Pew Research Center examined Americans' attitudes towards science and its role in politics.[9] Some areas of skepticism seemed to bridge the political divide. No more than twenty percent of Americans believe that scientists are transparent about potential conflicts of interest or trust scientists to admit their mistakes and take responsibility for them. On the other hand, sixty percent of Americans believe that scientists should play an active role in policy debates about scientific issues. The more fundamental divide appears, however, when it comes to scientists' role in politics and matters of value and judgment. According to Pew, "Americans are divided along party lines in terms of how they view the value and objectivity of scientists and their ability to act in the public interest." While most Democrats are likely to want experts involved in matters of policy and will trust their judgment, most Republicans "say scientists should focus on establishing sound scientific facts and stay out of such policy debates." Ultimately, the question that seems to divide people ideologically comes down to who should be entrusted with political power. The majority of Democrats responded that scientists are generally better at making decisions about policy that other people; while most Republicans "think scientists' [political] decisions are no different from or worse than other people's."[10]

It is that last issue that poses the existential issue about democracy and the true sovereignty of the People. Who should be entrusted with the power to rule: the People or the Experts? The answer to that question is in no way simple. Both sources of power have a tendency toward corruption. As Plato knew and as Socrates experienced, an elected legislature run on majority-rule can turn tyrannical just as easily as a King. Indeed, for hundreds of years before America's founding, the track record of democracies was abysmal. James Madison was haunted by the small republican experiments where dreams of democracy had quickly turned into nightmares of appalling tyranny. Recalling the Republic of Venice, he wrote that "[o]ne hundred and seventy-three despots would surely be as oppressive as one [king]."[11] On the other hand, many of the horrors of the Twentieth Century were accomplished at the cold hands of

---

[9] Pew Research Center Science & Society, "Trust and Mistrust in Americans' Views of Scientific Experts," Pewresearch.org, Accessed: May 28, 2021, https://www.pewresearch.org/science/2019/08/02/trust-and-mistrust-in-americans-views-of-scientific-experts/

[10] Pew Research Center Science & Society, "Trust and Mistrust in Americans' Views of Scientific Experts."

[11] A. Hamilton, J. Jay, and J. Madison, *The Federalist Papers*, 1788, at Federalist No. 48.

bureaucratic "experts." "[A]mong all the various political regimes and movements of the twentieth century, Communism, especially in its initial Soviet incarnation, happened to be the one most favorably predisposed toward science, believing most utterly...in science's power and value."[12] And yet, just as an example, it was centralized planning by "wise" Soviet economists and agricultural scientists that led to the *Holodomor* and the deaths of millions of Ukrainians in the 1930s.

The American Experiment

Arising out of the graveyard of failed experiments on both sides (rule by the People in democracies and rule by the Experts in oligarchies), the one undeniable success story has been the creation of the United States of America. The answer given by its founders is summarized in the large, bold letters of the first three words of the Constitution they wrote. It was their particular, historically unique, philosophy, combined with a novel feat of political engineering, that allowed America's founders to create a nation that is truly ruled by "We the People" while avoiding the historical tendency of democracies to devolve to tyranny.

Their philosophy was one based on individual liberty. Drawing from the greatest figures of the English Enlightenment, they believed that each person had a right to liberty that originated, not from any government, but from the law of nature itself. This right was inalienable. The entire purpose of government is just to *secure* the Blessings of Liberty. A legitimate government only enhances liberty, such as by providing fair, equal ground rules so that the People are freer—freer to associate, to enter contracts, to live life without interference from the evil or negligent actions of others, etc.

To accomplish this, the founders set out to engineer a structure for the nation's government that empowered it to create laws that secured liberty, while restraining it from enacting laws that would undermine individual liberty even if they were desired by the majority of people. The key stratagem they employed was genius. It was a type of political *jujitsu* as the founders channeled the worst, tyrannical instincts of the People and their representatives to ironically serve the interests of liberty. They wanted the various branches of government to be so busy fighting amongst themselves that they could not turn their tyrannous eye against the People. And they wanted every special interest, point of view, and political faction to have the freedom to be expressed frustratingly to the fullest, not in the form of law, but against the *countervailing* special interests, viewpoints

---

[12] A. Kojevnikov, "The Phenomenon of Soviet Science," *Osiris* 23, 1 (2008): pp.115-135.

and factions. The Constitutional structure they devised was meant to make it hard to pass laws. Debate among the country's factions was to be kept boiling, cancelling each other out for the sake of the people's liberty. Only a clear "winner" of an idea would become law, and that idea would necessarily not serve the interests of a faction, but rather the People as a whole and thereby preserve individual liberty as best as possible.

The problem is that the Constitutional structure that the Founders devised simply does not mix well with rule by experts. Delegating power to expert rule by scientific consensus throws oil onto the waters of constitutional democracy. The Constitution runs on debate over the best policy. Sure, science can help with the facts. But the Constitution anticipated that the domain of values, judgments, and policies would be left to free and healthy debate between factions whose special interest would, for the most part, cancel each other out. A claim that a certain policy is objectively, scientifically right does not bode well for debate and voting. If you think that your preferred policy proposal is as objectively "true" as 5+7=12 or the claim that the earth is round, you will think that debating your proposal is pretty foolish indeed.

The Administrative State

In fact, that is precisely what started to happen in America in the decades after the Civil War. Weary of never-ending and inefficient debate while the nation was rapidly industrializing, progressive reformers began to argue that there was a distinction between "politics" and "administration." Congress can endlessly struggle over the big ideas, but, they thought, let's have a corps of experts who can efficiently and scientifically "administer" Congress's basic policy prescriptions. For example, Congress can simply pass a law saying something uncontroversial like "we want fair competition," and a federal trade "commission" of experts will let us all know what that means scientifically by issuing regulations.

These administrative reformers of the late nineteenth and early twentieth century were completely candid and forthcoming about their disdain for the Constitutional structure. For them, the Constitution was either misguided from its start, or it was a relic of a bygone era that had little value for a complex, modern, and industrialized nation. Preeminent among them was the man who would become our 28th President, Woodrow Wilson. Before entering politics, Wilson was an academic. His scholarship claimed that America's founders had committed "the error of trying to do too much by vote." He believed that, compared to rule by Experts, the voice of the People is "meddlesome…a clumsy nuisance, a rustic handling delicate machinery."[13] "[A]n intelligent nation," according to Wilson "cannot be led or ruled save by thoroughly trained and completely-educated men" who have "special knowledge" and "disinterested ambition."[14]

Wilson's view eventually prevailed in the twentieth century, so much so that in 1948, political science professor Dwight Waldo memorably concluded that we now live in *The Administrative State* in his book of the same name. He identified the central conflict between the two sides' ideals of government and political power. According to his book, the disagreement was "between those whose hope for the future was primarily that of a planned and administered society, and those who, on the other hand, remained firm in the old liberal faith in an underlying harmony, which by natural and inevitable processes produce the greatest possible good."[15] Should we trust the People, properly constrained by the Constitutional structure, to exercise a sort of natural wisdom of the crowd? Or should we trust the Experts to use objective science in service of our collective march towards a greater society?

Outline

This book will examine that basic question—the one that divided Plato and Aristotle—from five different perspectives: American (chapter 2), historical (chapter 3), philosophical (chapter 4), scientific (chapter 5), and moral (chapter 6). And my answer will coincide with those first three words of our founders' Constitution.

America's founders believed deeply in the People and the value of self-rule. But they were realists about it. They would have concurred in Winston Churchill's observation that democracy is the worst form of government, except for all the others. They agreed with Aristotle that, if the danger of majoritarian tyranny and factional special interests can be properly controlled, the wisdom of the People will exceed even the wisest expert of administration. By examining the history of the progressive movement, it is clear that those that created and shaped America's administrative state were overtly opposed to the Constitution's founding principles. They made no bones about it. Rather than seeing rights as preceding government and originating in our status as individual human beings "endowed by our Creator" with inalienable liberty, they believed that it is the government itself that creates, provides, and protects

---

[13] W. Wilson, "The Study of Administration," *Political Science Quarterly* 2, 2 (1887): p.197.
[14] Quoted in R. Pestritto, "The Birth of the Administrative State: Where It Came From and What It Means for Limited Government," *First Principles Series* 16 (2007).
[15] D. Waldo, *The Administrative State* (The Ronald Press Co., 1948).

rights. Accordingly, such rights exist only to the extent that they serve the interests of the society, whose collective will is supposedly expressed through their government. Moreover, through study, they believed that experts and scientists can measure and predict collective outcomes of policy and administer government efficiently to optimize them.

Philosophically speaking, however, science is an excellent servant, but a terrible master. As the Scottish Enlightenment philosopher David Hume observed, science is equipped to tell us what "is," but it cannot deduce or infer what "ought" to be. That is exclusively the domain of moral values, a matter of debate of human judgments, opinions, and beliefs. Underlying any moral or political prescription that is supposedly based on science will be an unstated "metanarrative" based on values. This goes for the "right" side of the political spectrum as well as the "left." It is just as much true for "trickle-down economics" as it is for socialist programs and wealth redistributions. At some point, the scientist must be utilizing an underlying system of values to arrive at a policy prescription.

Accordingly, rule by Scientists must be based on the presupposition that individual scientists will have better systems of values, *i.e.*, better metanarratives, than the average person, or more accurately, than the People. Since science provides no special insights on questions of value, this position is nothing more than Plato's yearning for the philosopher-king. But by wrapping the philosopher-king in the garb of "science," an additional upshot can be had. The claim to "science" in policy debate will provide an unearned cudgel to the scientist-philosopher-king to quash debate and dissent as "anti-science." In governments that are purportedly democratic, it also allows elected politicians to hide behind the scientists for unpopular decisions, dismissing criticisms with the seemingly bullet-proof retort, "the science made me do it."

We will see that this is a problem because the People have a role to play in employing a critical eye to even the factual assertions of science. They need not be just passive receptors of indelible scientific truths. In the Seventeenth Century, the founder of modern empirical science, Sir Francis Bacon, believed that there were four "idols" that would tempt and distract experts from conducting true science. He called these the Idol of the Tribe, the Idol of the Cave, the Idol of the Marketplace, and the Idol of the Theater. Science may be as good as gold, but scientists are human and therefore fallible. Biases, undue deference to authority figures, career advancement and ambition, headline grabs, political or metaphysical preconceptions and prejudices, and the allure of statistical sleights-of-hand all threaten the objectivity of the scientific method. Since people are just as human now as they were in Bacon's time, these idols are still alive and well, and we will see numerous examples from both the hard and the social sciences. But in the final analysis, healthy skepticism, fair critique, and free and open debate are the Enlightenment's tools that the

People should be allowed to employ, not as a weapon against science, but as a fair and beneficial tool in its service.

Finally, as demonstrated by the examples of the eugenics movement and the Soviet *Holodomor*, a "scientific" administration of government tends toward inhumanity and a muting of the natural law as received by the individual moral consciences. Science has no choice but to view people as "objects." The ends of a "scientifically" administered government are objective: decreasing poverty, eliminating illness, increasing food production to feed the cities, etc. An ideology—whether from the political right or the political left—must always underlie a scientific administration. By contrast, America's founding principles are organized around the natural law. Each individual is treated as an end in him or herself, deserving of his or her own life, liberty, and dignity. Government is only legitimate to the extent it protects and enhances those things intrinsic to our humanity. The inhumanity of a scientific approach to government functions will be discussed with examples provided. Ironically, it can even be proven scientifically, through Stanley Milgram's famous experiments in psychology.

After examining these American, historical, philosophical, scientific, and moral topics, my proposed conclusion will be as straightforward as it is fundamental. To quote one of our greatest presidents, it is that America was "conceived in Liberty, and dedicated to the proposition that all men are created equal" and that, therefore, "government of the people, by the people, for the people, [should] not perish from the earth." The People are sovereign, and they can, should, and must rule themselves, keeping science in its worthy, important, and well-earned sphere as a fair arbiter of fact, not values.

Before proceeding, however, I will provide a few notes about how this book is written. Consistent with its democratic ideal, this book is meant to be by a person and for the people. For the most part, I have cited the "primary" sources where possible. You will see the real words of the Founding Fathers, the Progressives, the Enlightenment philosophers, the actual scientists whose studies are discussed, and other relevant historical figures. Indeed, this book is chock-full of direct quotations, often at some length. My hope is that this book's value will be primarily in how I have curated, organized, and marshalled this history and the thoughts and insights of these significant historical figures. Almost all of their writings are public domain and reviewing the full context of the quotations in these original works and/or historical artifacts is always an option. Ultimately, it does not really matter what I think about them or my subjective characterizations. You can make up your own minds after reading their actual words, and so I erred on the side of quotation rather than paraphrase or characterization.

Accordingly, my hope is that the style of writing employed by this book is a microcosm of its message. You have the power, and therefore, must be entrusted with the background "facts" to draw your own normative conclusion.

2.
Founding Principles

Whenever the political laws of the United States are to be discussed, it is with the doctrine of the sovereignty of the people that we must begin... In America the principle of the sovereignty of the people is not either barren or concealed, as it is with some other nations; it is recognized by the customs and proclaimed by the laws; it spreads freely, and arrives without impediment at its most remote consequences. If there be a country in the world where the doctrine of the sovereignty of the people can be fairly appreciated, where it can be studied in its application to the affairs of society, and where its dangers and its advantages may be foreseen, that country is assuredly America.

—Alexis de Tocqueville,
Democracy in America, vol. 1 (1835)

We the People.

It is easy to forget how radical the opening claim of our Constitution was. Sovereignty resides in individual people. And People come together to form a government to secure the Blessings of Liberty. Not to *provide* liberty. To *secure* it. This means that individual rights and freedoms precede government. Government exists only to protect and strengthen liberty. In fact, as America's other founding document—the Declaration of Independence—provides, it is the "Right of the People to alter or abolish" a government whenever it becomes "destructive" to the ends of liberty.[1]

The radical meaning of those first three words, written in large bold font, was not lost on those considering ratification of the Constitution. Patrick Henry believed that his State (Virginia) would better protect his rights than a large national government. He thought only a loose confederation of the thirteen states was appropriate to ensure the sovereignty of the people. On June 5, 1788, at the Virginia ratifying convention, Henry spoke in opposition to ratification of the proposed Constitution:

> The question turns, sir, on that poor little thing — the expression, We, the people, instead of the states, of America. I need not take much pains to show that the principles of this system are extremely pernicious,

---

[1] United States Declaration of Independence.

impolitic, and dangerous. Is this a monarchy, like England — a compact between prince and people, with checks on the former to secure the liberty of the latter?... You are not to inquire how your trade may be increased, nor how you are to become a great and powerful people, but how your liberties can be secured; for liberty ought to be the direct end of your government.²

In Massachusetts, Samuel Adams, brewer and leader of the Boston Tea Party, took similar umbrage at those first three words of the proposed Constitution:

I confess, as I enter the Building I stumble at the Threshold. I meet with a National Government, instead of a Federal Union of Sovereign States… [T]he Seeds of Aristocracy began to spring even before the Conclusion of our Struggle for the natural Rights of Men…So great is the Wickedness of some Men, & the stupid Servility of others, that one would be almost inclined to conclude that Communities cannot be free. The few haughty Families, think *They* must govern. The Body of the People tamely consent & submit to be their Slaves. This unravels the Mystery of Millions being enslaved by the few!³

The Constitution had been written against the backdrop of the failings of the Articles of Confederation. The Articles had failed to create a federal government strong enough to promote commerce and to instill the confidence necessary to attract domestic and foreign investments. Henry and Adams were opposed to the Constitution because they thought that it aimed only to increase the prosperity and greatness of the people, not their liberties. The individual states were already formed for to protect their citizens' liberty. How could a second government do anything but further impair liberty? According to them, the Constitution, therefore, ought only to be an undertaking of governments (states) because the people themselves may only form a government to secure liberty. Henry argued, "[t]he people have no right to enter into leagues, alliance, or confederations; they are not the proper agents for this purpose."⁴ It was a fair argument given the background beliefs about the government's purpose that the colonists generally shared—that a government, as opposed to a treaty

---

[2] Teaching American History, "Patrick Henry Speech Before Virginia Ratifying Convention - Teaching American History," Teachingamericanhistory.org, Accessed May 28, 2021, https://teachingamericanhistory.org/library/document/patrick-henry-virginia-ratifying-convention-va/

[3] H. Cushing, *The Writings of Samuel Adams* (New York: G.P. Putnam's Sons, 1904).

[4] Teaching American History, "Patrick Henry Speech Before Virginia Ratifying Convention - Teaching American History."

or confederation, can only exist to secure the people's inherent, pre-existing, and inalienable liberties.

Alexander Hamilton responded to arguments like those of Patrick Henry and Samuel Adams in the Federalist Papers, hoping to curry support for a federal government, predicated on an act of "We the People." Publishing under the pseudonym *Publius* (along with James Madison and John Jay), Hamilton argued against the need for a Bill of Rights in the Constitution because the structure of the Constitution ensured that the federal government would be one of the limited and enumerated rights that could, therefore, only *further secure* the People's liberties:

> Here, in strictness, the people surrender nothing; and as they retain every thing they have no need of particular reservations. "WE, THE PEOPLE of the United States, to secure the blessings of liberty to ourselves and our posterity, do ORDAIN and ESTABLISH this Constitution for the United States of America." Here is a better recognition of popular rights, than volumes of those aphorisms which make the principal figure in several of our State bills of rights, and which would sound much better in a treatise of ethics than in a constitution of government.[5]

But regardless of whether a founder thought that the Constitution would increase or decrease individual liberty, those on both sides of the founding debate agreed that protecting the liberty of the individually sovereign people was the sole purpose for forming a government. While they disagreed on the best mechanisms for accomplishing this, all of the founders believed in rule by the people. For example, among all of the founders, perhaps none of them favored a strong, central government more than Alexander Hamilton and John Adams. And yet, both of them explicitly rejected rule by expert administrators in the name of efficiency. Quoting a poem by Alexander Pope, Hamilton wrote that "we cannot acquiesce in the political heresy of the poet who says: 'For forms of government let fools contest, That which is best administered is best.'"[6] Alluding to the same verse the year the Declaration of Independence was signed, John Adams wrote that "Pope flattered tyrants too much." "Nothing could be more fallacious" than to claim that the best administered governments are best. Instead of administration, the best government is the one "whose principle and foundation is virtue."[7]

---

[5] A. Hamilton, J. Jay, and J. Madison, *The Federalist Papers*, 1788, at Federalist No. 84.
[6] Hamilton, Jay, and Madison, *The Federalist Papers*, at Federalist No. 68.
[7] Massachusetts Historical Society, "Adams Papers Digital Edition," Masshist.org, Accessed: May, 28, 2021, http://www.masshist.org/publications/adams-papers/index.php/view/PJA04dg2

No one in our founding era—not the staunchest federalist and certainly no anti-federalist—would have countenanced power coming to rest in the hands of even the best intentioned, best educated, wisest, and most efficient experts.

Ancient Greece

But it was not always believed to be so. The Enlightenment era in which America was founded was one-of-a-kind, a historical aberration. According to Alfred North Whitehead, "[t]he safest general characterization of the European philosophical tradition is that it consists of a series of footnotes to Plato."[8] The American experiment was one such footnote that departed markedly from Plato's more popular viewpoint from a historical perspective.

Despite his Athenian citizenship, Plato felt scorn for democratic forms of government. He had witnessed the unjust execution of his teacher, Socrates, at the hands of the direct Athenian democracy. In 399 B.C., a five-hundred-person Athenian jury convicted Socrates of corrupting the youth and impiety by a slim 56% majority, voting 280 to 220 to execute him. Years later, in his *Republic*, Plato assessed democracy as "a charming form of government, full of variety and disorder, and dispensing a sort of equality to equals and unequals alike."[9] The underlying premise of Plato's assessment is, of course, a far cry Jefferson's declaration—founded in Judeo-Christian and Enlightenment values—that "all men are created equal." Plato believed democracy would pass inexorably into tyranny and despotism. He concluded that democracy's fatal flaw was its yearning for liberty. "An excessive desire for liberty at the expense of everything else is what undermines democracy and [ironically] leads to the demand for tyranny."[10]

Plato analogized a nation to a ship and concluded that its leader or captain must be an expert. "The true navigator must study the seasons of the year, the sky, the stars, the winds, and all the other subjects appropriate to his profession if he is to be really fit to control the ship." He believed that a democracy is hostile to expert leadership because the People will "think that it's quite impossible to acquire the professional skill needed for such control and that there's no such thing as the art of navigation." So, instead, Plato concluded that true sovereignty must be vested in philosopher-kings:

> There will be no end to the troubles of states, or of humanity itself, until philosophers become kings in this world, or until those we now call

---

[8] A. Whitehead, D. Griffin, and D. Sherburne *Process and reality* (New York: Free Press, 1979).
[9] Plato and B. Jowett, *Plato: The Republic* (Charleston: Forgotten Books, 2008).
[10] Plato and B. Jowett, *Plato: The Republic*.

kings and rulers really and truly become philosophers, and political power and philosophy thus come into the same hands.[11]

For Plato, the ideal State, like the ideal man or the ideal ship, was an end in itself. It should be run like a well-oiled machine by those best disposed and trained to run and maintain it.

Plato's student, Aristotle, began to right the ship and recognized the nugget of gold at the core of Greek democracy at its best, as in Athens' golden age under Pericles' leadership. He believed that Plato's ideal Republic overvalued political unity, neglected the happiness of the individual, and conflicted with human nature.[12] Aristotle did, however, partially agree with Plato. Aristotle shared his concern regarding the propensity of democracy to lead to factionalism and injustice, and thence to tyranny. His ideal state, therefore, was a small city-state dependent on the prevalence of morally virtuous citizens in the middle class. Those that stand between rich and poor find it "easiest to obey the rule of reason," observed Aristotle.[13] "That the middle class is best is evident, for it is the freest from faction: where the middle class is numerous, there least occur factions and divisions among citizens."[14]

Responding to Plato's dream of the philosopher king, Aristotle theorized that the collective wisdom of the moderately-virtuous multitude would exceed the wisdom of any individual expert:

> The principle that the multitude ought to be supreme rather than the few best is one that is maintained, and, though not free from difficulty, yet seems to contain an element of truth....[I]f the people are not utterly degraded, although individually they may be worse judges than those who have special knowledge—as a body they are as good or better.[15]

If it could be pulled off—if democracies tendency to anarchy, rancor or tyranny could be controlled—Aristotle thought that democracy would be the best, or more accurately, the least bad form of government.

Monarchy and the Enlightenment

In the centuries after Greece and Rome fell, Plato's rule-by-the-few seemed to win out over Aristotle's view. Western civilization was generally governed under

---

[11] Plato and B. Jowett, *Plato: The Republic*.
[12] Aristotle and B. Jowett, trans., *Aristotle's politics and poetics* (New York: Viking Press, 1974), Politics II. 1-5.
[13] Aristotle and B. Jowett, trans., *Aristotle's politics and poetics*, Politics IV.11.
[14] Aristotle and B. Jowett, trans., *Aristotle's politics and poetics*, Politics IV.11.
[15] Aristotle and B. Jowett, trans., *Aristotle's politics and poetics*, Politics III. 11.

monarchies. Of course, whether these monarchs were, or could ever be, Plato's philosopher-kings was far from clear.

The cracks in feudal and monarchical systems of government started to form as the Age of Enlightenment began in the Seventeenth Century. Students of government and political theory began to question the theoretical basis for the hereditary monarchies that pervaded Europe. The traditional view—known as the "divine right of kings"—held that God had ordained that certain persons and their descendants ought to rule. But prominent British jurists like Sir Edward Coke argued against the idea that God vested sovereignty in kings, relying on the Magna Carta of 1215 as proof that the rights of Englishmen originated before William the Conqueror's invasion in 1066. This debate precipitated the English Civil War in 1642 between the Parliamentarians, led by Oliver Cromwell, and the Royalists. The Parliamentarians championed a proto-form of what America's founders later took for granted—that the legitimacy of any government is predicated on the consent of the governed. The political upheaval in England over the following decades culminated in Parliament's election of William and Mary as the new monarchs and in the enactment of the English Bill of Rights of 1689.

The liberties enumerated in the English Bill of Rights would sound very familiar to Americans today. They included the "freedom of speech," the "right of the subject to petition the king," the right of "Protestants" to "have arms for their defence," and the right to a jury trial, as well as prohibitions on "excessive bail" and "cruel and unusual punishments."[16] But perhaps even more important than the enumeration of these rights was the declaration of why they exist. The English Bill of Rights proclaimed that "the late King James the Second, by the assistance of divers evil counsellors, judges and ministers employed by him, did endeavour to subvert and extirpate the Protestant religion and the laws and liberties of this kingdom." Thus, the Bill was enacted "for the vindicating and asserting [the People's] ancient rights and liberties."[17] It is not too far of a cry from recognizing rights as "ancient" to seeing them as natural, God-given, and inalienable.

These political reforms were afforded their philosophical underpinnings by the English Enlightenment philosophers, including John Locke and David Hume. Locke, for example, alternatively titled his famous *Two Treatises of Government* as "In the Former, the False Principles and Foundation of Sir

---

[16] Avalon Project, "English Bill of Rights of 1689," Avalon.law.yale.edu, Accessed: May 28, 2021, https://avalon.law.yale.edu/17th_century/england.asp
[17] Avalon Project, "English Bill of Rights of 1689."

Robert Filmer, and His Followers, Are Detected and Overthrown. The Latter is an Essay Concerning the True Original, Extent, and End of Civil Government."[18] Accordingly, Locke's *First Treatise* is a near sentence-by-sentence refutation of Filmer's *Patriacha*, which espoused the theory of the divine right of kings. And his *Second Treatise* defended and supported the claim that all civil governments are charged by, and depend upon, the consent of the people.

It was on the shoulders of this history—ancient-Aristotelian and modern-Enlightenment—that America's Founders strove. The basic elements of liberty, of equality, and of "We the People" as the basis for government were all there, perhaps obliquely in Aristotle, but nearly entirely in Locke and the English Bill of Rights. Still, even if the *theory* was all in place, a modern and stable democratic republic had not yet been created in *practice* anywhere in the world. And that is where the true genius of the American Founders laid. They may not have been the scientists—the discovers—of democratic philosophy like John Locke. As the consummate practical Americans, they were its engineers.

Engineering Democracy

Plato had laid his finger on the potentially fatal flaw of democracy: the vice and selfishness of people that could devolve a democracy to a tyranny. For that reason, even Aristotle conceded that this required democracies to be relatively small and dependent on cultivating virtue among the people. He explained, "[t]o the size of states there is a limit, as there is to other things, plants, animals, implements, for none of these things retain their natural power when they are too large or too small."[19] This carried through to the instincts of Jefferson and Patrick Henry to prefer local, state governments and to seek to preserve the virtue of a hard-working agrarian middle-class. The Anti-Federalists (Henry chief among them) therefore championed Montesquieu who, echoing Aristotle in his *The Spirit of the Laws*, observed: "It is natural to a republic to have only a small territory, otherwise it cannot long subsist."[20]

To be sure, the Federalists agreed with the Anti-Federalists on the fallen nature of man and the dangers it posed to a democracy. Even the arch-Federalist, Alexander Hamilton, warned not to "forget that men are ambitious, vindictive,

---

[18] J. Locke, *Two treatises of government* (Dublin: William M'Kenzie, No. 33, College-Green, 1794).
[19] Aristotle and B. Jowett, trans., *Aristotle's politics and poetics*.
[20] C. Montesquieu, *The spirit of laws* (Glasgow: J. Duncan & Son, J. & M. Robertson, and J. & W. Shaw, 1793), ch. xvi. vol. I, book VIII.

and rapacious."[21] And James Madison referred to the poor history of even democracies with small territories as "among the most dark and degraded pictures which display the infirmities and depravities of the human character."[22] The Federalists were well aware that "[o]ne hundred and seventy-three despots would surely be as oppressive as one. Let those who doubt it, turn their eyes on the republic of Venice. As little will it avail us, that they are chosen by ourselves. An ELECTIVE DEPOSTISM was not the government we fought for."[23]

Throughout the Federalist Papers, they acknowledged that it was just a quixotic to hold out hope for a virtuous citizenry as the backbone of a democracy as it was for Plato to seek the virtuous philosopher-king. "If men were angels, no government would be necessary. If angels were to govern men, neither external nor internal controls on government would be necessary."[24] "A reverence for the laws would be sufficiently inculcated by the voice of an enlightened reason. But a nation of philosophers is as little to be expected as the philosophical race of kings wished for by Plato."[25] "Is it not time to awake from the deceitful dream of a golden age, and to adopt as a practical maxim for the direction of our political conduct that we, as well as the other inhabitants of the globe, are yet remote from the happy empire of perfect wisdom and perfect virtue?"[26]

Not even the almighty dollar could be relied upon for surety of good government in a democracy. Hamilton, for example, was a staunch supporter of capitalism and commerce if there ever was one. He had been deeply influenced by Adam Smith's *Wealth of Nations*. Years later, acting as Secretary of the Treasury, his *Report on Manufactures* to Congress drew extensively on Smith's work.[27] He even acknowledged that "the spirit of commerce has a tendency to soften the manners of men" and, through "mutual interest,…cultivate a spirit of mutual amity and concord."[28] He speculated that perhaps Shays Rebellion might not have occurred "[i]f Shays had not been a DESPERATE DEBTOR."[29] However, in

---

[21] Hamilton, Jay, and Madison, *The Federalist Papers*, at Federalist No. 6.
[22] Hamilton, Jay, and Madison, *The Federalist Papers*, at Federalist No. 37
[23] Hamilton, Jay, and Madison, *The Federalist Papers*, at Federalist No. 48
[24] Hamilton, Jay, and Madison, *The Federalist Papers*, at Federalist No. 51.
[25] Hamilton, Jay, and Madison, *The Federalist Papers*, at Federalist No. 49
[26] Hamilton, Jay, and Madison, *The Federalist Papers*, at Federalist No. 6
[27] E. Bourne, "Alexander Hamilton and Adam Smith," *The Quarterly Journal of Economics* 8, 3 (1894): p.328.
[28] Hamilton, Jay, and Madison, *The Federalist Papers*, at Federalist No. 6
[29] Hamilton, Jay, and Madison, *The Federalist Papers*, at Federalist No. 6.

light of numerous examples, Hamilton concluded that history had failed to demonstrate that commerce alone can maintain a democracy:

> Has it not, on the contrary, invariably been found that momentary passions, and immediate interest, have a more active and imperious control over human conduct than general or remote considerations of policy, utility, and justice?...Are not popular assemblies frequently subject to the impulses of rage, resentment, jealousy, avarice, and of other irregular and violent propensities?...Has commerce hitherto done anything more than change the objects of war?[30]

Something else needed to be relied upon to counteract the worst instincts of the people that lead to a degradation of democracy.

If it was not the size of a democracy, or the virtue of its citizens, or their mutual self-interest in commerce and trade that could make the great American experiment work, what was it? Hamilton pointed to the Constitution itself as the mean of "prevent[ing] the differences that neighborhood occasions, extinguishing that secret jealousy which disposes all states to aggrandize themselves as the expense of their neighbors."[31]

But to say the solution was the Constitution itself begged the question. And even more importantly, to say that democracy will work if everyone obeys the wise rules laid out in a constitutional document was really no answer either. The founders knew better than most that "mere demarcation on parchment of the constitutional limits of the several departments, is not a sufficient guard against those encroachments which lead to a tyrannical concentration of all the powers of government in the same hands."[32] Or as Supreme Court Justice Antonin Scalia would later put it in testimony before Congress:

> Every banana republic in the world has a bill of rights. Every president for life has a bill of rights. The bill of rights of the former evil empire, the Union of Soviet Socialist Republics, was much better than ours. I mean it literally. It was much better. We guarantee freedom of speech and of the press. Big deal. They guaranteed freedom of speech, of the press, of street demonstrations and protests, and anyone who is caught trying to suppress criticism of the government will be called to account. Whoa, that is wonderful stuff! Of course, just words on paper. What our Framers would have called a "parchment guarantee." And the reason is that the

---

[30] Hamilton, Jay, and Madison, *The Federalist Papers*, at Federalist No. 6
[31] Hamilton, Jay, and Madison, *The Federalist Papers*, at Federalist No. 6.
[32] Hamilton, Jay, and Madison, *The Federalist Papers*, at Federalist No. 48.

real constitution of the Soviet Union…did not prevent the centralization of power in one person or in one party. And when that happens, the game is over.[33]

No—the real genius of the Framers, their indelible feat of democratic engineering, was in turning the weaknesses of democracies into strengths that would paradoxically *serve* the People in the securing and preservation of their liberty. Large territories, diverse business interests, and vicious and self-interested citizens would be made to serve, rather than defeat, the republic. The cliché terms for this feat are "separation of powers" and "checks and balances." Indeed, most elementary students can recite the basics as described by James Madison in Federalist No. 51. Power is separated between three branches of government (legislative, executive, and judicial) so that the strivings of each respective branch prevent any of the others from gaining too much power and becoming despotic. This is all true, and the goal here is not to undersell the importance of these venerable doctrines one bit. But there is a subtle, unstated premise at play in these doctrines that is also crucial to Framer's design. It is one that operates not only across the branches of government but also intra-branch, for example, within the Congress, and among the citizens themselves in keeping them glued together as one nation. And it is the thing that will demonstrate that our Constitution is not meant for a government administered by scientists and other experts.

The Federalist Papers: Number 10

Until now, we have saved James Madison's Federalist No. 10. It may be the most important of the whole set of the Federalist Papers in expositing the genius of the Constitution and, for our purposes, the incompatibility of scientific rule with it. Madison's essay opens: "AMONG the numerous advantages promised by a well constructed Union, none deserves to be more accurately developed than its tendency to break and control the violence of faction."[34] After all, and as discussed above, a democracy's "propensity to this dangerous vice" of faction is its greatest weakness, even its "mortal disease under which popular governments everywhere [have] perished."[35]

---

[33] A. Scalia, "Opening Statement on American Exceptionalism to the Senate Judiciary Committee," Govinfo.gov, 2011, Accessed: May 28, 2021, https://www.govinfo.gov/content/pkg/CDOC-114sdoc12/pdf/CDOC-114sdoc12.pdf
[34] Hamilton, Jay, and Madison, *The Federalist Papers*, at Federalist No. 10.
[35] Hamilton, Jay, and Madison, *The Federalist Papers*, at Federalist No. 10.

Madison defined faction as a group of citizens, whether amounting to a majority or minority, "who are united and actuated by some common impulse of passion, or of interest, adverse to the rights of other citizens, or to the permanent and aggregate interests of the community." As Aristotle knew in emphasizing the need for a middle class, one of the most basic and deeper gulfs between people is that between the rich and the poor. Madison agreed: "the most common and durable source of factions has been the various and unequal distribution of property."[36] But, of course, there are other sources and causes of faction and even subdivisions within such faction: the religious and the non-religious, the farmer and the factory worker, the business person and the public servant, men and women, and the list goes on.

To address and engineer around this persistent and perennial danger, Federalist No. 10 proceeded analytically. "There are two methods of curing the mischiefs of faction: the one, by removing its causes; the other, by controlling its effects." Madison addressed and ultimately dismissed the first method, concluding that the causes of faction cannot be reasonably eliminated. He proposed that there are only "two methods of removing the causes of faction: the one, by destroying the liberty which is essential to its existence; the other, by giving to every citizen the same opinions, the same passions, and the same interests."[37] Of course, the first method could not be indulged—the entire purpose of government is the preservation of liberty. The most facile way to cure a disease may be to murder the patient, but then the entire purpose of medical science has been vitiated. Thus, destroying faction by destroying liberty was surely a remedy "worse than the disease," according to Madison.

The second method to eliminate the cause of faction—giving the people a unity of opinion—is "as impracticable as the first would be unwise" because "[t]he latent causes of faction are…sown in the nature of man." "The reason of man" is "fallible," which by necessity will lead to a difference of opinion. And even more importantly, a person's self-interest (as Madison put it, "his self-love") will inevitably influence his reason and lead to a difference of opinion. In other words, the goal of unifying opinion might be reduced to the goal of unifying the People's self-interests. However, Madison showed that equalizing self-interests would entail a destruction of liberty, and accordingly, could not be a viable option for the same reason as the first method of eliminating the cause of faction:

> The diversity in the faculties of men, from which the rights of property originate, is not less an insuperable obstacle to a uniformity of interests.

---

[36] Hamilton, Jay, and Madison, *The Federalist Papers*, at Federalist No. 10.
[37] Hamilton, Jay, and Madison, *The Federalist Papers*, at Federalist No. 10.

> The protection of these faculties is the first object of government. From the protection of different and unequal faculties of acquiring property, the possession of different degrees and kinds of property immediately results; and from the influence of these on the sentiments and views of the respective proprietors, ensues a division of the society into different interests and parties.[38]

Equality before the law does not necessarily entail equality of life outcomes. Madison argued that equality of outcome is not the proper goal of government. Only "[t]heoretic politicians…have erroneously supposed that by reducing mankind to a perfect equality in their political rights, they would, at the same time, be perfectly equalized and assimilated in their possessions, their opinions, and their passions."[39] Therefore, there is no way to eliminate the causes of faction in a democratic government. Or at least, any such government that succeeded in eliminating the causes of faction would not be a democratic one. The democratic engineers writing the Constitution needed to aim, instead, to control the *effects* of faction.

And it was in this goal that Madison believed the Constitution had succeeded. First, democracy will inherently prevent the undue influence of a minority faction because the majority will be enabled "to defeat its sinister views by regular vote. [The minority faction] may clog the administration, it may convulse the society; but it will be unable to execute and mask its violence under the forms of the Constitution." Therefore, the real danger is presented when a factional interest is back by a majority of the People and able to gain a majority stake in the government. In that instance, Madison rejected the idea that supposedly wise government officials would be a sufficient safeguard. No person is wise enough to weigh all the considerations to find the true and best path:

> It is in vain to say that enlightened statemen will be able to adjust these clashing interests, and render them all subservient to the public good. Enlightened statesmen will not always be at the helm. Nor, in many cases, can such an adjustment be made at all without taking into view indirect and remote considerations, which will rarely prevail over the immediate interest which one party may find in disregarding the rights of another or the good of the whole.[40]

---

[38] Hamilton, Jay, and Madison, *The Federalist Papers*, at Federalist No. 10.
[39] Hamilton, Jay, and Madison, *The Federalist Papers*, at Federalist No. 10.
[40] Hamilton, Jay, and Madison, *The Federalist Papers*, at Federalist No. 10.

Only the design of the government itself, not the virtue or expertise of the individuals operating it, can be trusted to rein in the tyranny of a majority—*the* problem that had vexed democracies since ancient times.

Accordingly, Madison proposed that America's government should be designed to prevent "the existence of the same passion or interest in a majority," or the majority "must be rendered, by their number and local situation, unable to concert and carry into effect schemes of oppression."[41] To that end, Madison rejected the collective wisdom of Aristotle, Montesquieu, and Patrick Henry, and held that it is better for a republic to be large. A large republic will be characterized by a vast diversity of interests, making it less likely that a majority will easily arise. Too, each representative in a large republic will represent a larger constituency, thereby making it "more difficult for unworthy candidates to practice with success the vicious arts by which elections are too often carried; and the suffrages of the people being freer, will be more likely to centre in men who possess the most attractive merit."[42] Size and diversity of viewpoint, which had once been thought to be the enemy of democracy, could actually be its savior if channeled wisely through the prism of government designed by the Constitution.

Rogue's Island and the Virtues of Difficult Lawmaking

As demonstrated by Federalist No. 10, Madison had learned well from the bad examples of small republics. Most notably and freshest in mind for Madison, were the abuses of the smallest state, Rhode Island, in the would-be United States. Rhode Island was so notorious in colonial America that, when the Constitution was being framed, many called it Rogues Island. For years, Rhode Island's legislature had engaged in unrestrained experiments with paper money, employing harsh measures to increase and force circulation of its currency. For example, the state's legislature passed an act providing that anyone refusing to take its money at face value would suffer a one-hundred pound fine for a first offense and might lose his rights as a citizen for a second.[43] When the state's judges found this act unconstitutional, the legislature fired them. Stories circulated of creditors in Rhode Island "leaping from rear windows of their houses, or hiding themselves in their attics" in order to escape debtors armed with the state's inflated paper money—waiting to make them an offer they literally couldn't refuse.[44] Rogues Island was so enamored with its small, utterly democratic, tyranny that it did not even bother sending delegates

---

[41] Hamilton, Jay, and Madison, *The Federalist Papers*, at Federalist No. 10.
[42] Hamilton, Jay, and Madison, *The Federalist Papers*, at Federalist No. 10.
[43] M. Jensen, *The new nation* (New York: Vintage Books, 1950), p. 324.
[44] C. Bullock, *Essays on the monetary history of the united states* (FORGOTTEN Books, 2015).

to the Constitutional Convention. It was the last of the former colonies to ratify the Constitution, and this was only after Congress threatened to pass a bill prohibiting all commercial intercourse with Rhode Island by the other states.

Thus, it was likely with Rhode Island's poor example in mind that Madison concluded that the wise engineer of a republic would:

> Extend the sphere [of the nation], and you take in a greater variety of parties and interests; you make it less probable that a majority of the whole will have a common motive to invade the rights of other citizens; or if such a common motive exists, it will be more difficult for all who feel it to discover their own strength, and to act in unison with each other... *A rage for paper money, for an abolition of debts, for an equal division of property, or for any other improper or wicked project, will be less apt to pervade the whole body of the Union than a particular member of it*[45]

The "extent and proper structure of the Union" was thus arranged by Madison to prevent a sufficient number of representatives from agreeing on anything based upon factional *interests*. Only those laws adapted to truly serve the "public weal" ought to command the assent of a sufficient number of officials in the various branches, each elected or appointed in different manners by different constituencies, to become law.[46]

To put it another way, the founders wanted it to be hard to cobble together a sufficient majority to create laws.

Enthusiasm for this design, inherent to the Constitution, was shared by founders beyond Madison. Hamilton, for example, alluded to Blackstone's famous formulation for a well-designed criminal justice system that "[i]t is better that ten guilty persons escape than that one innocent suffer."[47] So also, said Hamilton, it is better for good laws to not be passed than for a bad one to exert its terrible effects on the People:

> It may perhaps be said that the power of preventing bad laws includes that of preventing good ones; and may be used to the one purpose as well as to the other. ...[T]hose who can properly estimate the mischiefs of that inconstancy and mutability in the laws...will consider every institution calculated to restrain the excess of law-making, and to keep things in the same state in which they happen to be at any given period, as much more likely to do good than harm.... The injury which may possibly be done by

---

[45] Hamilton, Jay, and Madison, *The Federalist Papers*, at Federalist No. 10 (emphasis added).
[46] Hamilton, Jay, and Madison, *The Federalist Papers*, at Federalist No. 10.
[47] W. Blackstone and R. Kerr, *Commentaries on the laws of England* (London: J. Murray, 1880).

Founding Principles

defeating a few good laws, will be amply compensated by the advantage of preventing a number of bad ones.[48]

The Constitution was thus purposely designed to preserve liberty by ensuring that law had to originate from a *people* with fractured interests and goals, working through a government that was itself fractured and divided into departments, all making it very difficult for bad ideas to become law. This structure was fundamentally inconsistent with rule by experts. The last thing a philosopher-king wants is a structural barrier to the implementation of his or her eminently wise and infallible judgments. If one wanted to change the nature of the United States to be ruled by experts, the Constitution's restrained mechanic for law-making would need to be undermined. Accordingly, aspiring reformers favoring rule by experts have always leveled their aim directly between the eyes of Madison and Federalist Number 10.

For example, progressive historians like Charles Beard tried to tarnish Madison's motivations. They view the founding of the United States not as a triumph of Enlightenment values, but as a cynical attempt by the Founders to cement their own economic interests. Beard claimed that Federalist Number 10 shows that "the first and elemental concern" of the Constitution was "economic."[49] In other words, according to Beard, the "theories of government" that the Founders "entertain[ed] [were] emotional reactions to their property interests."[50] Madison supposedly betrayed his economic goals in Federalist Number 10. There, Beard claims, Madison admitted that he wanted the lower economic classes comprising a majority to fight among themselves and pose no threat to monied interest by being unable to get anything done. He argued that "the only way out [for Madison laid] in making it difficult for enough contending interests to fuse into a majority, and in balancing one over against another. The machinery for doing this is created by the new Constitution and by the Union."[51] Or, as another commentator put it, "What Madison prevents is not faction, but action. What he protects is not the common good but delay as such."[52]

Thus, the progressives sought to bring the debate full circle back to Plato and the need for an expert captain to lead efficiently and decisively. In the next

---

[48] Hamilton, Jay, and Madison, *The Federalist Papers*, at Federalist No. 73.
[49] C. A. Beard, *An Economic Interpretation of the Constitution of the United States* (New York: MacMillan, 1913).
[50] C. A. Beard, *An Economic Interpretation of the Constitution of the United States*.
[51] C. A. Beard, *An Economic Interpretation of the Constitution of the United States*.
[52] G. Wills, *Explaining America* (New York: Penguin Books, 2001).

chapter, we will see how and why these inroads against the Constitution's well-engineered democracy were made, starting in decades after the Civil War.

Conclusion

The Constitutional Convention occurred over the course of almost four months throughout the hot Philadelphia summer of 1787. The Founding Fathers believed that the Constitution they designed was truly novel. Democracies, republics, and even confederations of states were nothing new. The problem was that none had proven successful in maintaining a stable nation ruled by and for the People. They had either devolved into anarchistic bickering and infighting or had resulted in despotisms and monarchies like Rome.

If it was to succeed, the American difference was going to be a feat of engineering rather than pure political philosophy. Its beating heart was the idea of separation of powers and checks and balances at every imaginable level—between the federal government and the states, between the branches of government, and most importantly, between the majority of the people and the minority on any given issue. They knew that this was inefficient. And they knew that it was not a perfect solution. It had its costs. Some good laws were simply not going to be able to be passed. That cost, however, was justified by the strengthening of liberty and by the hope that the only laws that could be passed would be those that served the liberty of the People, rather than the interests of any faction or group. This was the summation and culmination of the Enlightenment's philosophy of government—the People come together to form a government only to secure and protect their liberties, not to surrender them in exchange for a tyrant's—or "expert's"—notion of the greater good.

Beard was wrong that the Founders were wolves in sheep's clothing, that behind their high-minded rhetoric and talk of "principles" was mere economic self-interest. No one can read the Founders papers and not hear the spirit of true belief and real excitement of new discovery and adventure.

The Constitutional Convention almost splintered and broke apart on numerous occasions. Alexander Hamilton left the Convention midway after one of his proposals was rejected for being too similar to the English monarchy. George Washington later wrote a letter to Hamilton about the course of the proceedings after his departure. He said: "I almost despair of seeing a favourable issue to the proceedings of the Convention, and do therefore repent having had any agency in the business."⁵³ The Convention was, in Washington's words, "mirac[ulously]"⁵⁴ saved time and time again by compromises that, rather than diminishing the quality of the final product as compromises so often do, actually improved it.

Indeed, after the Convention concluded, the Founders reported that they had felt that the hand of God had guided the proceedings and the ultimate design of the Constitutional government. Hamilton wrote an acquaintance, "For my own part, I sincerely esteem it a system which without the finger of God never could have been suggested and agreed upon by such a diversity of interests."⁵⁵ Madison believed that no individual could have designed the Constitution abstractly based upon pure political theory and philosophy. In Federalist Number 37, he questioned, "Would it be wonderful if, under the pressure of all these difficulties, the convention should have been forced into some deviations from that artificial structure and regular symmetry which an abstract view of the subject might lead an ingenious theorist to bestow on a Constitution planned in his closet or in his imagination?"⁵⁶ Instead:

> The real wonder is that so many difficulties should have been surmounted, and surmounted with a unanimity almost as unprecedented as it must have been unexpected. It is impossible for any man of candor to reflect on this circumstance without partaking of the astonishment. It is impossible for the man of pious reflection not to perceive in it a finger of that Almighty hand which has been so frequently and signally extended to our relief in the critical stages of the revolution.⁵⁷

Both Benjamin Franklin and George Washington summed up well the spirit of the Constitutional Convention viewed with the hindsight of its success. Franklin compared the new Americans to the ancient Israelites who had just been freed from Egypt. He believed that the Constitution, while not directly inspired like the law of Moses, was at least "in some degree influenced, guided, and governed by that omnipotent, omnipresent, and beneficent Ruler in

---

⁵³ Founders Online, "From George Washington to Alexander Hamilton, 10 July 1787," Founders.archives.gov, Accessed: May 28, 2021, https://founders.archives.gov/documents/Washington/04-05-02-0236
⁵⁴ Founders Online, "From George Washington to Alexander Hamilton, 10 July 1787."
⁵⁵ E. H. Scott, *Alexander Hamilton, John Jay, James Madison and Other Men of Their Time, The Federalist and Other Contemporary Papers on the Constitution of the United States* (New York: Scott, Foresman and Company, 1894), p. 646 (reproducing Alexander Hamilton to Mr. Childs, Wednesday, October 17, 1787).
⁵⁶ Hamilton, Jay, and Madison, *The Federalist Papers*, at Federalist No. 37.
⁵⁷ Hamilton, Jay, and Madison, *The Federalist Papers*, at Federalist No. 37.

Whom all inferior spirits live and move and have their being."⁵⁸ Franklin therefore warned Americans to trust the Constitution, and not to make the mistake the Israelites had of "suffer[ing] it to be worked upon by artful men, pretending public good."⁵⁹

Washington meanwhile extolled the new Constitution to his friend and long ally, the Marquis de Lafayette.⁶⁰ For Washington, it was "little short of a miracle, that the Delegates of so many different States…should unite in forming a system of national Government so little liable to well founded objections." The "two great points" or the "pivots on which the whole machine [of the new government] must move" and that supported his conclusion were: (1) that the new government's powers were limited; and (2) that those powers were "so distributed among the Legislative, Executive, and Judicial Branches, into which the general Government is arranged, that it can never be in danger of degenerating into a monarchy, an Oligarchy, an Aristocracy, or any other despotic or oppressive form, so long as there shall remain any virtue in the body of the People." In other words, they had succeeded in designing a Constitution with "more checks and barriers against the introduction of Tyranny…than any Government hitherto instituted among mortals hath possessed."⁶¹

Nevertheless, there will always be those that believe that the People should not and cannot rule themselves. They come in many guises, both on the right and on the left. Vigilance is always necessary because the Constitution is not self-executing. It needs the faith of the People to work; otherwise, it is just a piece of paper, a parchment guarantee not better than that of the U.S.S.R. After serving two terms as President under the Constitution, Washington advised the People of the need to "resist with care the spirit of innovation upon [the Constitution's] principles, however specious the pretexts" because "liberty itself [finds] its surest guardian" in the way the Constitution distributes and adjusts powers.⁶² He admonished "those entrusted with its administration to confine themselves within their respective constitutional spheres…A just estimate of [administrators'] love of power, and proneness to abuse it, which

---

⁵⁸ B. Franklin, *Comparing the Ancient Jews to the Antifederalists* (1788).
⁵⁹ B. Franklin, *Comparing the Ancient Jews to the Antifederalists*.
⁶⁰ Founders Online, "From George Washington to Lafayette, 7 February 1788," Founders.archives.gov, Accessed: May 28, 2021, https://founders.archives.gov/documents/Washington/04-06-02-0079
⁶¹ Founders Online, "From George Washington to Lafayette, 7 February 1788."
⁶² Our Documents, "Transcript of President George Washington's Farewell Address (1796)," Ourdocuments.gov, Accessed: May 28, 2021, https://www.ourdocuments.gov/doc.php?flash=false&doc=15&page=transcript

predominates in the human heart, is sufficient to satisfy us of the truth of this position."[63]

It is easy to want to abandon the fine-tuned machinery of the Constitution on the gamble that a philosopher king will rule wisely and justly. The work of democracy is hard. But, as Abraham Lincoln put it during his First Inaugural Address in 1861:

> Why should there not be a patient confidence in the ultimate justice of the people? Is there any better or equal hope in the world? In our present differences, is either party without faith of being in the right? If the Almighty Ruler of Nations, with His eternal truth and justice, be on your side of the North, or on yours of the South, that truth and that justice will surely prevail by the judgment of this great tribunal of the American people. By the frame of the Government under which we live this same people have wisely given their public servants but little power for mischief…[64]

---

[63] Our Documents, "Transcript of President George Washington's Farewell Address (1796)."
[64] Avalon Project, "First Inaugural Address of Abraham Lincoln, March 4, 1861," Avalon.law.yale.edu, Accessed: May 28, 2021, https://avalon.law.yale.edu/19th_century/lincoln1.asp

3.
The Rule of Scientists

What are the roots that clutch, what branches grow
Out of this stony rubbish? Son of man,
You cannot say, or guess, for you know only
A heap of broken images, where the sun beats,
And the dead tree gives no shelter, the cricket no relief,
And the dry stone no sound of water.

—T.S. Eliot, *The Wasteland*

Romanticism and the French Connection

Before considering how American political thought began to evolve after the Founding, it is helpful to compare the American Revolution with its historical mirror. The French Revolution is Bizarro to America's Superman. Belloq (in the film, a French archeologist) to America's Indiana Jones. As said Belloq to Jones, "Our methods have not differed as much as you pretend. I am but a shadowy reflection of you. It would take only a nudge to make you like me. To push you out of the light."[1] For our purposes, it will be illuminating to examine how the French Revolution was pushed out of the light and into the wasteland after the success of the American Revolution.

America's revolution occurred at the most fortuitous—or providential—of times. The signing of the Declaration in 1776 occurred at the very end of the Age of Enlightenment. While other factors were certainly at play, the baker's dozen of years that passed before France's Revolution in 1789 may have made a difference. In that short time, Romanticism had begun to displace Enlightenment thinking. The distinctions between Romanticism and Enlightenment are many and complex. But the primary one that helps make sense of the polar opposite results of the French and American Revolution is the Romantic's goal of elevating man.

The quintessential Romantic painting is Caspar David Friedrich's 1818 *Wanderer above the Sea of Fog*. It depicts a confident, individualistic man standing on the precipice of a craggy mountain, athwart and above a sea of fog and chaos in the forests below. The Romantics lionized Prometheus from Greek myth, who stole fire from the gods and was punished severely for it. The

---

[1] *Raiders of the Lost Ark*, Directed by S. Spielberg, Paramount Pictures (1981).

movement emphasized emotion, individuality, originality, and genius all in service to a prideful form of humanism.

Romanticism was a reaction to Enlightenment restraint and skepticism, especially about human nature. As established in the previous chapter, America's Founders were skeptical of human nature and the ability of individual experts to form sound political judgments by analyzing all relevant factors. Religious philosophers during the Enlightenment like Hamilton pointed to the fallen nature of man. But even atheist philosophers of that age like David Hume were deeply skeptical of man's epistemological abilities. He ended Book 1 of his *Treatise of Human Nature* with the conclusion:

> We have, therefore, no choice left but betwixt a false reason and none at all....The intense view of these manifold contradictions and imperfections in human reason has so wrought upon me, and heated my brain, that I am ready to reject all belief and reasoning, and can look upon no opinion even as more probable or likely than another.[2]

Hume extended this skepticism to the political sphere and was, at least in this respect, influential on America's Founders. He continued,

> When men submit to the authority of others, 'tis to procure themselves some security against the wickedness and injustice of men... But as this imperfection is inherent in human nature, we know that it must attend men in all their states and conditions; and that those, whom we chuse for rulers, do not immediately become of a superior nature to the rest of mankind, upon account of their superior power and authority... [Rulers will] be transported by their passions into all the excesses of cruelty and ambition.[3]

The echoes of Hume are found in the Federalist Papers' abiding suspicion of concentrated power in the hands of supposedly wise administrators.

Romantics outright rejected Hume's dreary humility about man's capabilities. Man, according to the Romantic, was destined to strive for intelligence, independence, and goodness. He is "heroic" in that old-fashioned, romantic sense. No only could he rule himself, he had the capacity to govern others virtuously, if only the corrupting influences holding him down could be swept aside.

---

[2] D. Hume, D. Norton, M. Norton, and D. Norton, *A treatise of human nature* (Oxford: Clarendon Press, 2014), Bk. 1, Section VII.
[3] Hume, Norton, Norton, and Norton, *A Treatise of Human Nature*, Bk. 3, Section IX.

It is no coincidence that the French philosopher Jean Jacques Rousseau is considered the Father of Romanticism[4] as well as the Father of the French Revolution. In his *Discourse on the Origin of Inequality*, Rousseau set out to prove "that man is naturally good."[5] It is only human society that "can have depraved him to such an extent" as to lead "men to hate each other in proportion as their interests clash." For example, according to Rousseau, the inventor of the concept of private property was the "real founder of civil society" and responsible for the "horrors and misfortunes" that ensued from not acknowledging that "the fruits of the earth belong to us all, and the earth itself to nobody."[6] But if the inherent nature of man is good, then, Rousseau thought, it must be possible to overcome the evil conditioning of society and rise above the chaos of faction and inequality. Rousseau proposed that this could be done by discovering the "General Will," which is the will of the People as a whole that must, by dint of man's benevolent nature, be aimed at the good. Accordingly, government should be organized by ministers tasked with discerning the General Will and implementing it. Only then would the corrupting influence of faction and individual interest be overcome through collective, social progress.

The revolutionaries toiling under the French monarchy were deeply influenced by Rousseau's views on the goodness of man in the state of nature, the inequality caused by the corrupting influence of society, and the project of instituting a government that could implement the General Will of the People. For centuries before the revolution, France was governed under the *Ancien Régime*. Like most of medieval Europe, the *Ancien Régime* was a system of hereditary monarchy coupled with a feudal system of French nobles. However, the explosion of France's population in the Eighteenth Century had led to economic depression, high unemployment and food prices, and increasing inequality. When the *Ancien Régime* failed to respond to these crises appropriately, it played directly into Rousseau's criticisms of civil society and monarchy.

The French revolutionaries acted as swiftly as the elites acted ineptly. Only a month after the *Ancien Régime* had convened an assembly to address the crisis, members of the assembly representing the peasant class broke away. In June 1789, the peasant representatives reassembled in the Tennis Court of Versailles and declared themselves the National Assembly. In what became known as the Tennis Court Oath, they vowed "not to separate, and to reassemble wherever circumstances require, until the constitution of the kingdom is established." The Bastille was stormed the following month on July 14. And by August 4, the

---

[4] R. Webb and J. Rousseau, *Jean Jacques Rousseau: The Father of Romanticism* (F. Watts, 1970).
[5] J. J. Rousseau, *Discourse On The Origin Of Inequality* (1754).
[6] J. J. Rousseau, *Discourse On The Origin Of Inequality*.

Assembly announced the abolishment of "the feudal system entirely," abrogating the special rights and privileges of the nobles and the clergy.

While the new "Legislative" Assembly tried to muddle its way through a sort of parliamentary constitutional monarchy with King Louis XVI over the following three years, the revolutionaries grew restless. The final blow was struck on August 10, 1792. The revolutionaries stormed the King's residence at the Tuileries Palace in Paris. And by the following month, in September 1792, the monarchy was formally abolished, and the Legislative Assembly was disbanded. The new French Republic was established, governed by the National Convention.

The Republic was governed by the National Convention and its successors until Napoleon ended the Republic and declared himself emperor of France a decade later. The National Convention was James Madison's worst nightmare from a checks-and-balances and separation-of-powers perspective. But it was paradise for a Rousseauian optimist, bullish on the power of the People to get things done in accord with the General Will. The National Convention was unicameral. It held plenary, unfettered, unenumerated legislative and executive powers, as well as judicial power over political cases. It tried, convicted, and executed the King within the first few months of its existence, and then established and oversaw judicial power through the Revolutionary Tribunal. While initially restrained by the moderate *Girondin* party, the Romantic spirit of the revolution in France was embodied more fully by the *Montagnards* (their apt name, meaning *mountain*, might spring to mind Friedrich's famous painting), who soon ascended to power.

Aside from the atrocities, the undertaking of the new French republic perhaps most indicative of its Romantic underpinnings was the institution of a new calendar. The year the Republic began (1792) became Year I in the new calendar. The twelve months were re-named based on nature in an attempt to erase the influence of the past. Fall months included the names *Brumaire* (mist) and *Frimaire* (frost). Spring included *Floreal* (flower). Summer had *Messidor* (harvest). Of course, the seven-day week of the Bible had to go and was replaced with much more "rational" and humanist ten-day weeks, named *primidi* (first day), *duodi* (second day), etc. The new calendar was adopted and implemented around the same time that *Notre Dame* and many other churches in France were converted into Temples of Reason. In November 1793 (or, better still, in *Brumaire* Year II), a Feast of Reason was held at *Notre Dame*, complete with an altar dedicated to Philosophy and Reason and with supplicants paying homage to an opera singer festooned as the Goddess of Liberty.

The *Montagnards* also went to work on re-writing the Republic's constitution, preserving the National Convention's stranglehold on all powers legislative, executive, and judicial. Comparing their *Declaration of the Rights of Man and of the Citizen* of 1793 with America's Bills of Rights is illuminating. According to the French declaration, equal eligibility for public employment is a "sacred and inalienable right."[7] Welfare (or "public relief") is a "sacred debt" because "[s]ociety owes maintenance to unfortunate citizens, either procuring work for them or in providing the means of existence for those who are unable to labor."[8] The right to education was also sacred and inalienable, and society must "put education at the door of every citizen."[9] The 1793 *Declaration* was meant to serve as an adjunct to the new *Montagnard* constitution of unlimited centralized power. But even the plans for that constitution were abandoned when the *Montagnards* determined that further consolidation of emergency powers to "defend the Revolution" were required.

The leader of the *Montagnards* and the *Jacobin* political club from which they arose was Maximilien Robespierre. When the *Montagnards* determined that further emergency powers were needed, he formed the Committee of Public Safety and led France into the notorious Reign of Terror. A tribunal of philosopher-kings that would make Plato proud, the Committee's efforts included conducting war, investigating, prosecuting, and executing dissidents and counter-revolutionaries, and enforcing central economic planning. For example, the Committee's Law of General Maximum was instituted in 1793 in response to food shortages. It set price limits on food and punished price gouging. Hoarding of grain became a capital offense, and many thousands of French citizens were executed by guillotine for political crimes or otherwise running afoul of the Committee.

Robespierre deeply admired Rousseau and had fully imbibed the revolutionary spirit of the time. He was the author of France's famous republican motto: *Liberté, Égalité, Fraternité*. He was devoted to the revolution in every fiber of his being and was nicknamed *L'Incorruptible* because of his reliable zeal and steadfast refusal to accept bribes. As one might expect, a person like Robespierre was exceptionally frank in his speeches and writings, especially when he was at the height of his power in 1793 and 1794. These materials provide candid insights into the philosophical mindset underlying the French Revolution to underscore the comparison with America's Founders, especially as it relates to the rule by the experts.

First, Robespierre held an exalted opinion of man's capabilities. At the 1793 Festival of the Supreme Being, he exhorted: "Man, whoever you might be, you can yet conceive high thoughts on your own; you can attach your fleeting life

---

[7] Columbia University, "Declaration Rights of Man 1793," Columbia.edu, Accessed: May 28, 2021, http://www.columbia.edu/~iw6/docs/dec1793.html
[8] Columbia University, "Declaration Rights of Man 1793."
[9] Columbia University, "Declaration Rights of Man 1793."

to God Himself and immortality."[10] And he practically quoted Rousseau in a 1794 political essay, proclaiming: "Happily, virtue is natural in the people, [despite] aristocratical prejudices."[11]

Second, Robespierre believed that government, in implementing the General Will of a good people, should aspire to the same heights as man. His confidence in the revolution and the ability of the revolutionaries to employ logic and reason to promote the general welfare leaps off the page of his writings:

> What is the first object of society? It is to maintain the inviolable rights of man. What is the first of these rights? The right to exist....In accordance with this principle, what is the problem to be resolved in the matter of legislation on subsistence? It is this: to assure to all members of society the enjoyment of the portion of the fruits of the earth that is necessary to their existence...It's claimed that they are impractical; I say that they are as simple as they are infallible. *It is claimed that they offer an insoluble problem, even for those of genius; I say that they present no difficulty to good sense and good faith.*[12]

Third, in light of man's and government's unbounded potential, Robespierre demanded that his government take decisive action through the hands of wise officials and administrators. He rallied them, saying:

> [W]e have more to fear from the excesses of weakness, than from excesses of energy. The warmth of zeal is not perhaps the most dangerous rock that we have to avoid; but rather that languor which ease produces and a distrust of our own courage. Therefore continually wind up the sacred spring of republican government, instead of letting it run down....[T]he wisdom of government should guide its operations according to circumstances...Terror is only justice prompt, severe and inflexible; it is then an emanation of virtue; it is less a distinct principle than a natural consequence of the general principle of democracy, applied to the most pressing wants of the country.[13]

The political constraints that America's founders poured their sweat into and debated passionately, like separation of powers and respect for minority viewpoints, were worthless to Robespierre. To Robespierre, the Federalist Papers would have seemed counterproductive and weak pearl-clutching.

---

[10] M. Robespierre, *Speech at 1793 Festival of the Supreme Being* (1793).
[11] M. Robespierre, *On the Principles of Political Morality* (1794).
[12] M. Robespierre, *On Subsistence Goods* (1792) (emphasis added).
[13] M. Robespierre, *On the Principles of Political Morality* (1794).

Humility must take a back-seat to progressive vigor. And so, according to Robespierre, "everywhere we must level the obstacles and hindrances to the execution of the wisest measures" of government.[14] The Committee on Public Safety should be empowered to act as philosopher-kings. He defiantly defended the Committee's extraordinary powers, asking "Do you believe that without unity in action, without secrecy in [the Committee's] operations, without the certainty of finding support within the Convention that the government could triumph over so many obstacles and so many enemies? No."[15]

It is hard to imagine a more diametric difference in the approach to a republic than that of Robespierre and *Publius* (the pen-name of Alexander Hamilton, James Madison, and John Jay in the Federalist Papers). *Publius* was suspicious of power, tremulous in constitutional design, skeptical of man's ability to resist corruption. Robespierre had moved beyond the Enlightenment. He was a true Rousseauian and a true Romantic, seeing himself at the precipice of the *Montagnard*, conqueror of the fog and chaos of his superstitious ancestors.

And in a quintessential Romantic irony, the Frankenstein's monster Robespierre created eventually turned around on its creator. Robespierre was arrested, convicted, and guillotined over the span of three days in July 1794 (or, as Robespierre would have undoubtedly preferred, *Thermidor* Year II). While the moderates who assumed power after Robespierre's execution attempted to control the government and right the revolutionary ship, Napoleon overthrew the Republic within five years of Robespierre's death, and declared himself emperor for life within ten.

During the first throes of the French Revolution, it might have been difficult to discern how it differed from America's. Thomas Jefferson was slow to recognize the danger inherent to the French Revolution's approach to democracy. Jefferson initially commended the French National Assembly's "firmness and wisdom" and professed the "highest confidence" in its ability to govern.[16] Though with the benefit of hindsight in Jefferson's late-in-life correspondence with John Adams, he later conceded that the French Revolution failed at its own misguided hands

---

[14] M. Robespierre, *For the Defense of the Committee of Public Safety* (1793).
[15] M. Robespierre, *For the Defense of the Committee of Public Safety* (1793).
[16] Founders Online, "From Thomas Jefferson to Diodati, 3 August 1789," Founders. archives.gov, Accessed: May 28, 2021, https://founders.archives.gov/documents/Jefferson/01-15-02-0317

and was "defeated by Robespierre."[17] He regarded as "folly" the newer generation's (the Romantics') preference for "self-learning...of rejecting the knowledge acquired in past ages, and starting on the new ground of intuition."[18]

If Jefferson's life had not been so invested in the idea of revolution, he might have discerned the French difference more easily. Indeed, the dispassionate assessment of Irish statements Edmund Burke in the early days of the French Revolution proved more prescient than Jefferson. Burke, who had supported the American Revolution, immediately recognized the dangerous ideas at play in France. He published his *Reflections on the Revolution in France* in 1790, well before the King Louis was executed and before Robespierre's rise to power.

Burke agreed with the approach employed in America, whose framers had designed their constitution like cautious engineers. For Burke, the "science of government" needed to be "practical" and required "infinite caution" and "experience, and even more experience than any person can gain in his whole life, however sagacious and observing he may be."[19] Burke contrasted this humble caution with the "theorists" of France who tended to "extreme[]" ideas, which "in proportion as they are metaphysically true, they are morally and politically false." Under the abstract generalizations of Rousseauian politics,

> [a]ll the pleasing illusions which made power gentle and obedience liberal, which harmonized the different shades of life, and which, by a bland assimilation, incorporated into politics the sentiments which beautify and soften private society, are to be dissolved by this new conquering empire of light and reason.[20]

Burke proved prophetic about the end result of the French Resolution, the human cost of which was made clear by Robespierre's confident frolic into terror. Burke summarized their error best: "They have no respect for the wisdom of others, but they pay it off by a very full measure of confidence in their own....Their liberty is not liberal. Their science is presumptuous ignorance. Their humanity is savage and brutal."[21]

---

[17] Founders Online, "From Thomas Jefferson to John Adams, 4 September 1823," Founders. archives.gov, Accessed: May 28, 2021, https://founders.archives.gov/documents/Jefferson/98-01-02-3737

[18] Founders Online, "Thomas Jefferson to John Adams, 5 July 1814," Founders.archives.gov, Accessed: May 28, 2021, https://founders.archives.gov/documents/Jefferson/03-07-02-0341

[19] E. Burke, *Reflections on the Revolution in France* (1790).

[20] E. Burke, *Reflections on the Revolution in France*.

[21] E. Burke, *Reflections on the Revolution in France*.

The Foundations of American Progressivism

Unlike the French Revolution, the tendency for unrestrained democracy to devolve to savagery and brutality was admirably resisted by the American Constitution's sound structural engineering, as discussed in the previous chapter. For more than one hundred years after America's founding, the republic was ruled by the People for the most part in accordance with the restraining influence of the Founders' designs. Of course, this was not without its costs. For decades, the government of the Union was unable to eliminate the scourge of slavery due, at least in part, to the checks on the majority built into the governmental structures. One could easily see a Robespierre ending slavery with the stroke of a pen and perhaps a few strokes of the guillotine. But in the United States, abolitionists found themselves boxed into compromises and half-measures. The Northwest Ordinance promised that freedom for all would be the rule in first significant expansion of the nation's territory. And the Missouri Compromise helped to maintain the Union in its early days while limiting the expansion of further slave territories. And yet, the evil institution of slavery stubbornly persisted.

The blotch of slavery was the exception that otherwise proved the rule of good governance under the Framer's designs. It may even be argued that the persistence of slavery, and the need for the Civil War to eliminate it, did not reflect a flaw in the Constitutional structure of government, but rather a failure in realizing the foundational basis of the Constitution in natural law and human liberty preceding government. The engineering of the Constitution only works if the promise of the Declaration of Independence—that "all men are created equal [and] endowed by their Creator with certain unalienable Rights [including] Life, Liberty, and the pursuit of Happiness"—undergirds it.[22]

Indeed, even before the adoption of the Constitution, the laws of some states recognized that the freedom of the People—all people—was a prerequisite to a functioning republic. One well-reported case involved the status of a person named Quork Walker in Massachusetts.[23] A Massachusetts resident named Nathaniel Jennison claimed to own Mr. Walker as a slave based on a bill of sale given in 1754 from Zedekiah Stone when Walker was nine months old. In 1783, Mr. Walker brought a civil indictment against Jennison alleging that he was assaulted when Jennison undertook to restrain him. The entire Massachusetts Supreme Judicial Court presided over the trial of his case, entitled Walker v. Jennison. Chief Justice William Cushing's well-reasoned instructions to the jury, which enabled Mr. Walker to prevail and win his freedom with a verdict finding Jennison guilty, have been preserved. He told the jury:

---

[22] United States Declaration of Independence.
[23] A. Rugg, "William Cushing," *The Yale Law Journal* 30, 2 (1920), p. 132-33.

> [Contrary to the practices in Europe], a different idea has taken place with the people of America, more favorable to the natural right of mankind, and to that natural, innate desire of Liberty, with which Heaven, (without regard to color, complexion or shape of noses, features) has inspired all the human race. And upon this ground our Constitution of Government, by which the people of this Commonwealth have solemnly bound themselves, sets out with declaring all men are born free and equal—and that every subject is entitled to liberty,—and to have it guarded by the laws, as well as life and property—and in short is totally repugnant to the idea of being born slaves. This being the case I think the idea of slavery is inconsistent with our own conduct and Constitution; and there can be no such thing as perpetual servitude of a rational creature, unless his liberty is forfeited by criminal conduct or given up by personal consent or contract.[24]

Chief Justice Cushing's reasoning was, obviously, sound and proper. If the United States Supreme Court had ruled similarly in the infamous case of *Dred Scott v. Sandford*,[25] the Civil War may even have been avoided. But ultimately, it was the People themselves that ended the abomination of slavery—through their blood in the War and with their votes in the duly enacted Thirteenth Amendment to the Constitution. No expert or philosopher king ended slavery or preserved the Union. The People did.

Nonetheless, the post-war world had changed since the time of the founding, and the erosion of Enlightenment values in American had finally begun. Even before the Civil War, American philosophers—most preeminent among them, Ralph Waldo Emerson—had begun to absorb the spirit of Romanticism. In a watershed moment, Emerson delivered his oration entitled *The American Scholar* before the Harvard Phi Beta Kappa Society in 1837. There, Emerson lamented America's "sluggard intellect," hidden under "its iron lids" by the "exertions of mechanical skill," that had not yet fulfilled the "expectation[s] of the world."[26] Like Rousseau's General Will, Emerson championed the "One Man" emanating from "the whole society." "Man is not a farmer, or a professor, or an engineer, but he is all. Man is priest, and scholar, and statesman, and producer, and soldier." Imagining the tragedy of the "One Man" being "spilled into drops," Emerson longed to cure man of his "degenerate state" exemplified by his instinct to pursue individual goals and interests as a "victim of society."

---

[24] Rugg, "William Cushing," p. 132-33.
[25] *Dred Scott v. Sandford*, 60 U.S. (19 How.) 393 (1857).
[26] R. W. Emerson, *The American Scholar* (1837).

Like Robespierre, Emerson exuded confidence and vivacity regarding man's place in the world:

> The one thing in the world, of value, is the active soul....The soul active sees absolute truth; and utters truth, or creates. In this action, it is genius…In its essence, it is progressive. The book, the college, the school of art, the institution of any kind, stop with some past utterance of genius. This is good, say they, – let us hold by this. They pin me down. They look backward and not forward. But genius looks forward: the eyes of man are set in his forehead, not in his hindhead: man hopes: genius creates.[27]

It "becomes" the American Scholar, concluded Emerson, to "feel all confidence in himself," regardless of "[s]ome great decorum, some fetish of a government."[28] Fortunately, Emerson's transcendentalism led him to prefer a limited government and a more individualized approach to manifesting the One Man in the world. He did not frequently delve into the issues of day-to-day politics. But Emerson proved a highly influential American philosopher, and the seed of his Romanticism influenced others more politically inclined, who became known as Progressives in the decades that followed.

Progressives sardonically dubbed the years after the Civil War and Reconstruction, the Gilded Age. But whether it was a Golden Age or merely Gilded was a matter of perspective. The Gilded Age was a period of vast and rapid economic growth and industrialization. The United States became more interconnected through competition in railroads and new communication technologies. America's immense need for labor was filled by millions of immigrants seeking economic opportunity and freedom. As expected in a free market, the average wealth of the American citizen increased significantly in this period. In 1880, the average annual wage for an industrial worker was $380. By 1890, it rose almost 50% to $564 per year.[29] The prevalence of child labor declined.[30] The United States economy grew faster during these years than during any other time in all of its history.[31]

Believing that these economic advances were only skin-deep gilding, the Progressives focused on wealth inequalities and perceived injustices. Even if

---

[27] Emerson, *The American Scholar*.
[28] Emerson, *The American Scholar*.
[29] United States Census Office, "1890 Census Bulletin (11th Census), Statistics of Manufactures," October 22, 1892, p.2.
[30] Ibid.
[31] E. Kirkland, *Industry comes of age: Business, Labor, and Public Policy* (New York: Holt, Rinehart and Winston, 1961), pp. 400–05.

the total pie was getting much bigger, larger percentages of that wealth concentrated into smaller percentages of the population. After all, these were the years of John D. Rockefeller, Andrew Carnegie, Cornelius Vanderbilt and other railroad "tycoons," and the fabled "robber baron." Substandard and dangerous working conditions led to anger, strikes, and unionization. The story was the same then as it is today. And the key conclusion drawn from the premises of injustice and inequality was that the political system in America had become corrupt to favor the rich industrialist.

While not a household name today, Henry George was an early and very popular social and economic commentator responsible for spreading socialist solutions to Gilded Age inequality and suffering. He put a perceptibly American spin, with religious undertones, on typical European socialist rhetoric. His landmark 1879 treatise, *Progress and Poverty: An Inquiry into the Cause of Industrial Depressions and of Increase of Want with Increase of Wealth: The Remedy*, sold millions of copies in the years after it was published. For a time, it eclipsed the sales of all other books, except the Bible. In the work, George posited that industrialization had proven itself to lack any "tendency to extirpate poverty or to lighten the burdens of those compelled to toil. It simply widens the gulf between Dives and Lazarus, and makes the struggle for existence more intense."[32] Like socialist and communist philosophers throughout the ages, George focused on relative inequality rather than the absolute advances made by each social strata during the Gilded Age. He observed that "[i]t is true that wealth has been greatly increased, and that the average of comfort, leisure, and refinement has been raised; but these gains are not general." Accordingly, the free market "should be spurned by the statesman, scouted by the masses, and relegated in the opinion of many educated and thinking men to the rank of a pseudo science." Instead, George believed, the "methods of political economy" needed to be employed to eradicate the unfairness of inequality.[33]

Specifically, George thought he had "traced the unequal distribution of wealth which is the curse and menace of modern civilization to the institution of private property in land." And so his solution to the problem was, in a literal sense, communism:

---

[32] H. George, *Progress and Poverty: An Inquiry into the Cause of Industrial Depressions and of Increase of Want with Increase of Wealth: The Remedy* (London: W. Reeves, 1884).
[33] George, *Progress and Poverty: An Inquiry into the Cause of Industrial Depressions and of Increase of Want with Increase of Wealth: The Remedy*.

> To extirpate poverty, to make wages what justice commands they should be, the full earnings of the laborer, we must therefore substitute for the individual ownership of land a common ownership. Nothing else will go to the cause of the evil—in nothing else is there the slightest hope.[34]

George acknowledged that it might not be right, or consistent with the American ethos, for the government to physically seize all land from its current owners. So, to functionally accomplish this, he instead proposed the elimination of all taxes, except that the government would impose a 100% tax on all rents of land.

George's solution to inequality was clearly radical. Many would think that it was decidedly un-American, and his ultimate policy prescription on rents obviously did not come to pass. But the immense popularity of his work showed that he had placed his finger on a tender spot in the culture and the American mind of that time. Many agreed that the unequal situation of the masses was intolerable and that the free market had proven unsuited to the problems of the modern age. Yet, in place of the sweeping changes to the economic system proposed by George, Progressives with a canny eye for politics saw that the real opportunities lay instead in altering the machinery of American government to accomplish such economic ends indirectly.

Administrative Progressivism

The famous French tourist and commentator on America, Alexis de Tocqueville, recognized the danger that centralized administration posed to American democracy, especially under the influence of the more European models. "The democratic nations of Europe have all the general and permanent tendencies which urge the Americans to centralization of government."[35] In Europe, this transformation had begun when "the charitable establishments of Europe" which were "formerly in the hands of private persons or of corporations" became "dependent on the supreme government" and "in many countries...actually administered by that power."[36]

> The State almost exclusively undertakes to supply bread to the hungry, assistance of shelter to the sick, work to the idle, and to act as the sole reliever of all kinds of misery. Education, as well as charity, is become in most countries at the present day a national concern. The State receives,

---

[34] George, *Progress and Poverty: An Inquiry into the Cause of Industrial Depressions and of Increase of Want with Increase of Wealth: The Remedy*.
[35] A. De Tocqueville, *Democracy in America* (London: Penguin Books, 2003).
[36] De Tocqueville, *Democracy in America*.

and often takes, the child from the arms of the mother, to hand it over to official agents: the State undertakes to train the heart and to instruct the mind of each generation.[37]

Tocqueville's key insight in comparing European social programs with American charity was this: in Europe, government undertook this role to indirectly accomplish the ends it truly wanted. Tocqueville saw that government oversight and administration of more and more areas of its citizens' daily life were not necessarily undertaken so that the administrator could "over-zealous[ly]…settle[] points of doctrine."[38] It was not to directly persuade the citizenry of, for example, Henry George's political points. Rather, Tocqueville warned, these responsibilities were happily undertaken so that the administrator could obtain "more and more hold upon the will of those by whom doctrines are expounded." For example, by depriving the clergy of their property and paying their salaries, the governmental administrators "make them their own ministers—often their own servants—and by this alliance with religion they reach the inner depths of the soul of man." For that reason, "[a]ll governments of Europe have in our time singularly improved the science of administration: they do more things…[and] they seem to be constantly enriched by all the experience of which they have stripped private persons."[39] Thus, if American Progressives were to make similar inroads on the hearts and minds of the people, their efforts too should become focused on expanding the scope of so-called "administrative law."

To overcome the restraints of democratic faction and individualism that categorized the founders, America found its champion of administration in the academy. In the 1880s, the future president of the United States, Woodrow Wilson, was a graduate student at Johns Hopkins University. At twenty-nine years old and before graduating with his Ph.D., he published his first significant political work entitled *Congressional Government* in 1885. There, Wilson exhibited scorn for the Framer's core insight in designing America's government with the goal of separation of powers. While acknowledging that it was the "main purpose" of the Constitutional Convention, Wilson believed that the Constitution suffered from a "radical defect" in that "it parcels out power and confuses responsibility as it does."[40] To Wilson, "checks and balances" have "proved mischievous just to the extent to which they have succeeded in establishing themselves as realities." For that reason, the "charm of our

---

[37] De Tocqueville, *Democracy in America*.
[38] De Tocqueville, *Democracy in America*.
[39] De Tocqueville, *Democracy in America*.
[40] W. Wilson, *Congressional Government* (1885).

constitutional ideal" had worn off on him, and he felt no obligation to "honor[]" the Constitution with "blind worship." To take its place, government needed some "magistrate or representative body" that "can decide at once and with conclusive authority what shall be done...immediately." Much like Robespierre, Wilson saw one "principle" of government "clearer" than any other: that "*somebody must be trusted*" with efficient power.[41] And for that reason, Wilson's early work touted a parliamentary form of government where the parliament exerts direct control over ministers of the law, and he argued that "civil service reform" in the United States was necessary to bring us closer to that paradigm.

Soon after receiving his doctorate, Wilson expounded on his ideas for the civil service in the highly influential essay entitled *The Study of Administration*.[42] At the age of thirty-one, Wilson wondered why practitioners of political "science" had heretofore "thought, argued, dogmatized only about the *constitution* of government." The more interesting question for Wilson was "how the law should be administered with enlightenment, with equity, with speed, and without friction."[43] Of course, this was nothing more than a magic trick; Wilson's claim that there is a distinction at all between politics and administration was mere misdirection. The Founders knew well that the constitutional structures they were designing were meant, in themselves, to dictate the mechanism of administration. Administration under the Constitution was meant to be fraught with politics; it was meant to be democratic. Wilson begged the question in assuming that, whatever their methods of politics, governments should be administered with "speed" and without "friction." The Founders would have said that their government was designed to slow things down to allow the frictions of democracy to work themselves out.

Anticipating that the Founders' philosophy would be brought to bear against his own, Wilson replied that the Constitutional structures of the Founders were simply a product of their time. The Founders thought "[t]he functions of government were simple, because life itself was simple."[44] Back then, "[t]here was no complex system of public revenues and public debts to puzzle financiers." The complexity of the modern era demanded a new "science" of administration. The government should be "everywhere putting its hands to new undertakings." For example, "[t]he utility, cheapness, and success of the

---

[41] Wilson, *Congressional Government*.
[42] W. Wilson, "The Study of Administration," *Political Science Quarterly* 2, 2 (1887): p.197.
[43] Wilson, "The Study of Administration," p.197.
[44] Wilson, "The Study of Administration," p.197.

government's postal service...point towards the early establishment of governmental control of the telegraph system." Wilson felt inspired by the "governments of Europe" in running telegraph and railroad enterprises, and felt that "no one can doubt that in some way [our government] must make itself master of masterful corporations."[45]

But where, Wilson wondered, should we look to improve this new "science of administration"? After all, "American writers have hitherto taken no very important part in the advancement of this science." The English and the Americans had focused on studying "the art of curbing executive power to the constant neglect of the act of perfecting executive methods. It has exercised itself much more in controlling than in energizing government." And so, Wilson looked elsewhere, to the European models. The science of administration "has found its doctors in Europe...It has been developed by French and German professors." You see, Wilson observed, the Europeans had identified the problem for us Americans, and it was "popular sovereignty." It is inefficient that "[i]n order to make any advance at all we must instruct and persuade a multitudinous monarch called public opinion." The "People" have a "score of differing opinions" and "can agree upon nothing simple," much less anything complex, Wilson thought. "Wherever regard for public opinion is a first principle of government, practical reform must be slow and all reform must be full of compromises."[46] Instead of the People, the Continental theorists and Wilson fervently desired a "civil service" of philosopher kings to rule.

Wilson believed that this civil service—his army of administrators—should be cloaked in immense power and discretion. Paradoxically (and naively), he thought that "the greater [the bureaucrat's] power, the less likely is he to abuse it, the more is he nerved and sobered and elevated by it." Moreover, Wilson's bureaucrats should not be held to account by the People. The Founders had committed "the error of trying to do too much by vote." Public criticism of bureaucrats is "meddlesome...a clumsy nuisance, a rustic handling delicate machinery." This system of administration envisioned by Wilson "will require not a little wisdom, knowledge, and experience." The "important and delicate extension of administrative functions" needs to be accomplished by that ubiquitous feature of the administration state: the "Commission." The service of the state must be "removed from the common political life of the people....Its motives, its objects, its policy, its standards, must be bureaucratic."[47]

---

[45] Wilson, "The Study of Administration," p.197.
[46] Wilson, "The Study of Administration," p.197.
[47] Wilson, "The Study of Administration," p.197.

While Wilson's *The Study of Administration* had proposed an amiable divorce between politics and administration, it did not take long for Wilson's hostility to the politics of democracy to take him further. In 1891, he argued that "[t]he functions of government are in a very real sense independent of legislation, and even constitutions... Administration cannot wait upon legislation, but must be given leave, *or take it*, to proceed without a specific warrant in giving effect to the characteristic life of the State."[48] Like most Progressives, Wilson exhibited the confidence that "an intelligent nation cannot be led or ruled save by thoroughly trained and completely-educated men" who have "special knowledge" and "disinterested ambition."[49]

While chasing these theoretic pursuits, Wilson's career trajectory seemed to embody his academic yearnings for the rule of well-educated philosopher kings like him. He quickly climbed the political ladder. In 1902, he became the President of Princeton University. Eight years later, he was elected the governor of New Jersey. When Teddy Roosevelt's Bull Moose Party split the Republican vote with the incumbent William Howard Taft, Wilson won the 1912 Election for the Presidency of the United States. And while Wilson focused on his ascent to political power, his academic mantle in the emerging field of administrative law was taken up by successors such as Frank Goodnow.

Both Goodnow and Wilson served as president of the American Political Science Association in the years after its founding in 1903. Goodnow spent his early years in the academy teaching administrative law at Columbia, and eventually became President of Johns Hopkins University, Wilson's alma mater, in 1914. The influence of Wilson's purported distinction between politics and administration is easily discerned in Goodnow's early work: *Politics and Administration: A Study in Government*, published in 1900.[50] Like Wilson, Goodnow expressed frustration with prior academics' insistence on studying the Constitution and its history. The change in the times required government to adapt through the creation of an administrative state, the discharge of whose functions "must be uninfluenced by political considerations." Again like Wilson, Goodnow admired the European mindset of administrative deference and believed that "until the people of the United States attain to the same conception" as the Europeans, America's political branches "cannot be intrusted" with administrative power. Goodnow believed that administration of the law through

---

[48] R. Pestritto, "The Birth of the Administrative State: Where It Came From and What It Means for Limited Government," *First Principles Series* 16 (2007).
[49] Ibid.
[50] F. Goodnow, F. *Politics and Administration: A Study in Government* (New York: MacMillan Co., 1900).

the political branches of government would "pollute the sources of truth" and lead to "bias" and "corruption."[51]

Administrators and bureaucrats, on the other hand, were merely disinterested scientists according to Goodnow. The job of administrative bureaucrats is to "attend[] to the scientific, technical, and, so to speak, commercial activities of government."[52] They must be "free from the influence of politics" because "their mission is the exercise of foresight and discretion, the pursuit of truth, the gathering of information, the maintenance of a strictly impartial attitude toward the individuals with whom they have dealing, and the provision of the most efficient possible administrative organization." In acting as supposed experts, Goodnow believed that administrative agencies were well-suited to "express the will of the state as to details where it is inconvenient for the legislature to act."[53] Of course, Goodnow's breathtaking assertions betrayed the influence of Rousseau in the notion that a "General Will" of the People can be discovered scientifically, as well as the specter of Robespierre's impatience with legislative inconvenience.

Most importantly, Goodnow's proposal in *Politics and Administration* is deeply and irreconcilably at odds with the Founder's Constitution. And Goodnow certainly knew it. And so, Goodnow began taking direct aim at the Founders in his follow-on works, aptly titled *Social Reform and the Constitution* and *The American Conception of Liberty and Government*.

In *Social Reform*, Goodnow echoed Wilson's complaint about Americans' "reverence" for the Constitution, which he felt was "superstitious."[54] Even if the Constitution was fine for its time in the Eighteenth Century, modernization had left it to "working harm rather than good" because of the need for swift, efficient, and scientific administrative reform. The Founder's high-flying rhetoric about "natural law," popular sovereignty, and inalienable rights was "worse than useless" according to Goodnow because they only served to "retard development."[55] Instead of "fundamental principles of universal application" favoring liberty, Goodnow preferred the German, more specifically, Hegelian conception that a society (and its government) must evolve like a living organism, a collective progressing through history as one sacred unit.

---

[51] Goodnow, F. *Politics and Administration: A Study in Government*.
[52] Goodnow, F. *Politics and Administration: A Study in Government*.
[53] Goodnow, F. *Politics and Administration: A Study in Government*.
[54] F. Goodnow, *Social reform and the constitution* (New York: Franklin, 1911).
[55] Ibid.

In *The American Conception of Liberty and Government*, Goodnow rebuffed the philosophical underpinnings of America's other founding document: the Declaration of Independence. He derided Jefferson's idea that people were born with "certain unalienable rights" as pure "speculation" and lacking any "historical justification."[56] For him, it is "impossible to assert, as a matter of fact even, that man existed first as an individual and that later he became…a member of human society." Instead, he lauded the post-Enlightenment view from Europe that people have no "inherent rights" but only "so much as rights which find their origin in the law as adopted by" their government.

> The rights which he possesses are…conferred upon him, not by his Creator, but rather by the society to which he belongs. What they are is to be determined by the legislative authority in view of the needs of that society.[57]

Goodnow believed that Americans had only misguidedly clutched at their flawed conceptions of liberty for so long because of "the religious and moral influences in this country" that "emphasized personal responsibility and the salvation of the individual soul" leading to "unadulterated individualism."[58] Truly chilling stuff, to be sure, and yet they were written by Goodnow as the president of a major American university. These ideas, laying at the heart of what would become Progressivism, became mainstream in the early Twentieth Century.

Simply put, Goodnow believed that Americans had now outgrown the idea that government should be "organized primarily, if not exclusively, for the purpose of securing liberty," and were now mature enough to realize that it should be organized "to secure social efficiency instead."[59] "A great many questions of policy, particularly those which have to do with governmental activities, cannot be intelligently initiated by those who have had no administrative experience."[60] Therefore, America ought to be led by "the expert service" of the "learned professions" trained in diverse disciplines, such as law, medicine, engineering, education, agriculture, journalism, and business administration.

But there was a final hurdle to be overcome. Even if "expert" progressives could gain political power (as they would with the election of Wilson),

---

[56] F. Goodnow, *The American conception of liberty and government* (Providence: Standard Printing Company, 1916).
[57] Goodnow, *The American conception of liberty and government*.
[58] Goodnow, *The American conception of liberty and government*.
[59] Goodnow, *The American conception of liberty and government*.
[60] Goodnow, *The American conception of liberty and government*.

Goodnow saw that America's courts were the final bulwark that needed to be overcome. The courts were, in his opinion, too "much under the influence of the political philosophy of the eighteenth century."[61] Courts had gone wrong in believing that the Constitution was "superior" to the acts of a duly elected legislature, causing courts to strike down laws when they trammel upon the Constitutional rights reflecting the natural rights of the People.[62]

Accordingly, in his later work, Goodnow wondered:

> What methods there are by which pressure may be brought to bear upon the courts to induce them either to abandon or not to adopt the conception that our constitutions postulate a fixed and unchangeable political system and a rigid and inflexible rule of private right, and to apply the rule that constitutions, which are practically unamendable, should be considered rather as statements of general principles whose detailed application should take account of changing conditions, and should be so interpreted by judicial decision as to be susceptible of a continuous and uninterrupted development.[63]

Thus, the quest to cajole the courts to endorse the idea of the living, evolving Constitution had begun.

The Courts Bless the Administrative State

In his two terms, Wilson did what he could to advance the progressive agenda. Before World War I side-lined of his domestic agenda, Wilson oversaw the implementation of the first federal income tax, created the Federal Reserve System to centralize bank administration, pushed Congress to pass the Adamson Act to establish the eight-hour workday enforced with overtime pay, and created the Federal Trade Commission to enforce the new Clayton Antitrust Act. During the War, Wilson appointed the Creel Committee (also known as the Committee on Public Information) to influence public opinion about the war effort through propaganda. Concurring in Goodnow's view that rights are whatever a society finds expedient for the time, Wilson also urged Congress to pass the Sedition Act of 1918 to criminalize speech critical of the government. And once the War was over, Wilson famously called for the establishment of the League of Nations, the forerunner of the United Nations.

But when Wilson left office in 1921, the pendulum of public opinion had swung the other way and the Progressive Era came to an end, at least until the

---

[61] Goodnow, *The American conception of liberty and government*.
[62] Goodnow, *The American conception of liberty and government*.
[63] F. Goodnow, *Social reform and the constitution* (New York: Franklin, 1911).

Great Depression. The 1920s saw a return to smaller government under the leadership of Warren G. Harding and Calvin Coolidge. Coolidge stands out in history as a defender of America's founding principles. He summarized his key insights and opposition to progressivism during a speech he delivered on the 150th anniversary of the signing of the Declaration of Independence. Coolidge affirmed that the "people have to bear their own responsibility" for the ideals of self-government.[64] "There is no method by which that burden can be shifted to the government." He believed that further progress beyond America's ideals was a false promise and ultimately, regressive:

> If all men are created equal, that is final. If they are endowed with inalienable rights, that is final. If governments derive their just powers from the consent of the governed, that is final. No advance, no progress can be made beyond these propositions. If anyone wishes to deny their truth or their soundness, the only direction in which he can proceed historically is not forward, but backward toward the time when there was no equality, no rights of the individual, no rule of the people. Those who wish to proceed in that direction can not lay claim to progress. They are reactionary. Their ideas are not more modern, but more ancient, than those of the Revolutionary fathers.[65]

The Declaration "is the product of the spiritual insight of the people. We live in an age of science and of abounding accumulation of material things. These did not create our Declaration. Our Declaration created them. The things of the spirit come first."[66]

In addition to the presidential administration, the final bulwark of liberty under the United States Supreme Court was continuing to hold in opposition to the growing administrative state. In 1921, Harding appointed former President William Howard Taft to be the Chief Justice of the United States Supreme Court (the only person to have served in both offices). Taft led the Court for less than a decade before his death in 1930. But in that time, his most important contribution may have been in the burgeoning area of administrative law. Most notably, in 1926, Taft wrote the Court's opinion in the case of *Myers v. United*

---

[64] Teaching American History, "Speech on the 150th Anniversary of the Declaration of Independence," Teachingamericanhistory.org, Accessed May 28, 2021, https://teachingamericanhistory.org/library/document/speech-on-the-occasion-of-the-one-hundred-and-fiftieth-anniversary-of-the-declaration-of-independence/

[65] Teaching American History, "Speech on the 150th Anniversary of the Declaration of Independence."

[66] Teaching American History, "Speech on the 150th Anniversary of the Declaration of Independence."

States,[67] which he would later claim was his most important opinion while he served on the Court.[68]

While legally important, the facts of the case dealt with a simple dispute over a very provincial bureaucracy. Woodrow Wilson, the great champion of tenure for public servants, had appointed Frank Myers as the postmaster for Portland for a four-year term, but fired him less than three years later. Myers sued claiming that his termination violated the Tenure of Office Act, which restricted a President's power to fire certain administrative office holders. Writing for the Supreme Court's majority, Taft ruled against Myers. He held that the tenure act was unconstitutional because it impeded the President's oversight of administrators. Under the Constitution, such administrators must be accountable to the People through presidential oversight. Taft's seventy-page opinion masterfully relied upon and discussed the principles of the country's Founders. Page after page of his opinion is consumed with quotations from Madison and Hamilton, the Constitutional Convention, the Federalist Papers, the states' ratification conventions, and the debates of the First Congress. Taft based his ruling on fundamental values of liberty protected by the "maintenance of independence, as between the legislative, executive and the judicial branches" to provide "security for the people." And so "those in charge of and responsible for administering functions of government" by virtue of their due election to office under the Constitution must have power over those appointed to assist them.[69]

Taft knew that his decision was drawing a line in the sand against the growth of the administrative state. He observed that, in the modern era,

> Laws are often passed with specific provision for adoption of regulations by a department or bureau head to make the law workable and effective. The ability and judgment manifested by the official thus empowered, as well as his energy and stimulation of his subordinates, are subjects which the President must consider and supervise in his administrative control.[70]

No matter the administrator's claim to disinterested and scientific expertise, the division between politics and administration must be curtailed. According to Taft and the Founders, the administrator must be held accountable to the President and thereby, to the People.

After *Myers*, the Supreme Court also resisted efforts whereby Congress's laws themselves were becoming unaccountable to the People in favor of expert

---

[67] *Myers v. United States*, 272 U.S. 52 (1926).
[68] W. H. Taft, Letter to W. M. Bullitt, Nov. 4, 1926.
[69] *Myers v. United States*, 272 U.S. 52 (1926).
[70] *Myers v. United States*, 272 U.S. 52 (1926).

administrators. Once the Great Depression began, Progressives rose to power again as people looked to government for solutions to the great economic problems of the time. Franklin Delano Roosevelt was elected in 1932 to implement his "New Deal" with the American People. One component of Roosevelt's program was the National Industrial Recovery Act passed by Congress at Roosevelt's request in 1933 (NIRA). The Act created the National Recovery Administration, to which Congress delegated vast discretion to regulate American industry to achieve *fair* wages and prices and reduce "destructive competition."

The A.L.A. Schechter Poultry Corporation was indicted under NIRA for violating the Administration's "Live Poultry Code." In order to promote what it thought was "fair competition," the Code required chicken wholesalers to sell only entire coops of chickens to retailers and butchers. A.L.A. Schechter, however, had allowed New York butchers to choose which chickens they wanted to buy without also being required to buy the other chickens from the same coop. After it was convicted of violating the NRA's Code, the company sought review from the United States Supreme Court. And in its opinion in *A.L.A. Schechter Poultry Corporation v. United States*,[71] the Court found NIRA, the National Recovery Administration, and its "codes" of competition unconstitutional. It explained that "Congress is not permitted to abdicate or to transfer to others the essential legislative functions" that the Constitution vests in Congress itself.[72] To fulfill the constitutional structure and ensure accountability to the People, the policies and frameworks for law must originate with Congress, not unelected experts or trade groups.

The *Schechter Poultry* decision was issued in 1935, just as President Roosevelt's frustration with the Supreme Court was starting to boil over. It would not be long before Roosevelt would embark on a plan to pack the Supreme Court to ensure that his legislative agenda would be enforced. But in 1935, the Supreme Court was already starting to soften.

On the same day, *Schechter Poultry* was issued, the Supreme Court eroded Chief Justice Taft's *Myers* opinion—less than a decade old at that point—with its judgment in *Humphrey's Executor v. United States*.[73] President Hoover had appointed William Humphrey to serve as a Federal Trade Commissioner in 1931. When President Roosevelt took office, he wanted the FTC to more aggressively regulate and police supposedly "unfair methods of competition."

---

[71] *A.L.A. Schechter Poultry Corporation v. United States*, 295 U.S. 495 (1935).
[72] *A.L.A. Schechter Poultry Corporation v. United States*, 295 U.S. 495 (1935).
[73] *Humphrey's Executor v. United States*, 295 U.S. 602 (1935).

Humphrey disagreed and wanted the FTC to take a more *laissez-faire* approach. Roosevelt wrote to Humphrey, asking him to resign his commission, saying "You will, I know, realize that I do not feel that your mind and my mind go along together on either the policies or the administering of the Federal Trade Commission."[74] Humphrey refused to resign because Congress had provided that commissioners could serve their full term unless removed by the President "for inefficiency, neglect of duty, or malfeasance in office."[75] After Roosevelt forcibly removed Humphrey from office in October 1933, Humphrey sued. Ironically, President Roosevelt relied on *Myers* for his authority to unilaterally remove Humphrey for any reason. But this time, the Supreme Court disagreed, holding that Roosevelt had wrongfully terminated Humphrey.

Unlike *Myers*, absent from the Court's fourteen-page opinion was any in-depth discussion of the Founding era evidence and the principles underlying separation of powers. The Court thought that rehashing the history in which Taft so thoroughly steeped his opinion "would add little of value to the wealth of material" collected in *Myers* to decide the "narrow point...that the President had power to remove a postmaster."[76] It acknowledged that some of Taft's discussion contained broader "expressions...which tend to sustain [the President's] contention," but "[i]nsofar as they are out of harmony with the views here set forth, these expressions are disapproved." The Court wanted to rush past the history and change direction from Taft's more originalist predispositions.

Instead, the Court focused its efforts in discussing how the FTC was meant to be "nonpartisan" and "impartial[]." Its arguments could have been easily drawn directly from the academic writings of Wilson or Goodnow:

> [The FTC] is charged with the enforcement of no policy except the policy of the law. Its duties are neither political nor executive, but predominantly *quasi*-judicial and *quasi*-legislative. Like the Interstate Commerce Commission, its members are called upon to exercise the trained judgment of a body of experts appointed by law and informed by experience...[O]ne advantage which the commission possessed...lay in the fact of its independence, and that it was essential that the commission should not be open to the suspicion of partisan direction...[T]he commission was not to be subject to anybody in government...free from political domination or control...separate and

---

[74] *Humphrey's Executor v. United States*, 295 U.S. 602 (1935).
[75] *Humphrey's Executor v. United States*, 295 U.S. 602 (1935).
[76] *Humphrey's Executor v. United States*, 295 U.S. 602 (1935).

apart from any existing department of the government [and] not subject to the orders of the President.[77]

In other words, it was important for the FTC to "act in discharge of their duties independently of executive control," which Congress had accomplished by fixing the terms of service for the commissioners.[78]

Two decades later, Justice Robert Jackson would concede the judicial legerdemain inherent in that little word "*quasi*" that made all the difference to the Court in *Humphrey's Executor*:

> Administrative agencies have been called quasi-legislative, quasi-executive, or quasi-judicial, as the occasion required, in order to validate their functions within the separation of powers scheme of the Constitution. The mere retreat to the qualifying "quasi" is implicit with confession that all recognized classifications have broken down, and "quasi" is a smooth cover which we draw over our confusion, as we might use a counterpane to conceal a disordered bed.[79]

Nevertheless, the years and decades that followed *Humphrey's Executor* saw the creation of myriad "independent" agencies, all supposedly justified by their expertise and scientific acumen to solve on-the-ground problems supposedly ill-suited to Congress, the President, or the federal courts. In 1935 alone, Congress created the National Labor Relations Board and the Social Security Administration. Another leap forward occurred under President Johnson, where, among others, the National Transportation Safety Board and the Occupational Health and Safety Review Commission were born. Independent commissions were tasked with overseeing a vast portion of the legal landscape, including the Post Office, consumer product safety, nuclear energy, labor relations, and Indian gaming.

In the late 1970s and 1980s, Congress began targeting core functions of the executive and judicial branches for "expert" control. For example, it had long been thought that criminal prosecution was a core executive function that should be directly controlled by the President. But Congress still created an office for an "independent counsel" to investigate and prosecute certain types of corruption within the government, which the Supreme Court found constitutional in *Morrison v. Olson*.[80] Fixing the sentences for defendants

---

[77] *Humphrey's Executor v. United States*, 295 U.S. 602 (1935).
[78] *Humphrey's Executor v. United States*, 295 U.S. 602 (1935).
[79] *Federal Trade Commission v. Ruberoid Co.*, 343 U.S. 470, 487-88 (1952).
[80] *Morrison v. Olson*, 487 U.S. 654 (1988).

convicted of crimes had long been thought a core judicial function that should be directly controlled by the courts. And yet, Congress created the United States Sentencing Commission to promulgate binding sentencing guidelines to govern how federal judges issued their sentences. Again, the Supreme Court found this framework constitutional in *Mistretta v. United States*.[81]

Justice Antonin Scalia dissented in *Morrison* and *Mistretta*. He understood how it was all too easy for politicians to pass the buck to "the experts" to fulfill their democratic responsibilities. Now, with the Supreme Court's blessing over the middle decades of the Twentieth Century, he foresaw:

> all manner of 'expert' bodies, insulated from the political process, to which Congress will delegate various portions of its lawmaking responsibility. How tempting to create an expert Medical Commission (mostly M.D.'s, with perhaps a few Ph.D.'s in moral philosophy) to dispose of such thorny, 'now-in' political issues as the withholding of life-support systems in federally funded hospitals, or the use of fetal tissue for research.[82]

In this way, the administrative state was born and grew after all three branches of government—Congress, the President, and the Supreme Court—failed to act jealously to protect their domains of power as the Founders had hoped they would.

Conclusion

The words of the founders of America's administrative state have been extensively quoted because they show that they did not try to hide or obfuscate their underlying philosophical commitments and principles. Their perspective was fundamentally at odds with those who framed the Constitution. They knew it. And they wanted you to know it too. They outright professed that the founding principles were relative, a product of their time, and had become outdated in a modern era of industrialization.

The progressive politicians and political scientists of the late-nineteenth and early twentieth centuries were dissimilar to those of today. With a few exceptions, most today try to pay lip service to the Founders and the Constitution, either not understanding their philosophy or actively hiding their disdain for it. But whether then or now, they all buck the Enlightenment's political insight that individual rights are grounded in natural law, which precedes human law. If that axiom is granted, then a democratic government

---

[81] *Mistretta v. United States*, 488 U.S. 361 (1989).
[82] *Mistretta v. United States*, 488 U.S. 361 (1989) (Scalia, J. *dissenting*).

must find a way to ensure that laws are a product of popular sovereignty—that they find their basis in We the People, the consent of the governed. James Madison and the other delegates of the Constitutional Convention believed that they had found a way to do this. Their work explained how they had coopted the weaknesses of democracy—faction and size—into strengths to serve the interests of liberty and the rights of the People. The problem is that Progressives found they did not trust the People and hence rejected Madison's democratic solutions. They rejected Aristotle's claim that a certain wisdom exists in the People exceeding the wisdom of any individual.

They questioned how the People could understand a complex, modern, industrialized society, even when they act through representatives. And more to the point, even if representatives understand the problems of such a society, their solutions to those problems will inevitably be infected by self-interest, by faction, and by the inefficiencies of the democratic process. The same enlightenment that had given birth to democratic theories of government had also given birth to modern science. And so, Progressives turned away from the People and to Science. Woodrow Wilson called it a turn away from the study of "politics" to the study of "administration." To them, the country needed "experts" to rule through their wisdom unfettered by politicians who would only slow them down. Democracy should be boiled down to the People's choice to get the experts involved. All that was needed from the People was to hire the experts to fix the problem.

John Dewey was one of the most, if not the most, prominent American philosophers of the early twentieth century. Like Wilson, he received his doctorate from Johns Hopkins University. He made broad contributions in many fields of philosophy, including epistemology, logic, psychology, religion, and politics. In the same year that the Supreme Court decided *Humphrey's Executor* (1935), Dewey published *Liberalism and Social Action*, which summarized the basis for Progressives' attack on America's founding principles. As the title suggests, Dewey believed that liberalism—the ideas of individual liberty and popular sovereignty embodied by the Constitution—had proven itself incapable of undertaking the social action necessary in the modern world.

Dewey was a relativist and criticized what he saw as the founder's arrogance in attempting to set down the eternal principles of government.

> The earlier liberals [i.e., the Founders] lacked historic sense and interest. For a while this lack had an immediate pragmatic value. It gave liberals a powerful weapon in their fight with reactionaries. For it enabled them to undercut the appeal to origin, precedent and past history by which the opponents of social change gave sacrosanct quality to existing inequities

and abuses. But disregard of history took its revenge. It blinded the eyes of liberals to the fact that their own special interpretations of liberty, individuality and intelligence were themselves historically conditioned, and were relevant only to their own time. They put forward their ideas as immutable truths good at all times and places; they had no idea of historic relativity, either in general or in its application to themselves.[83]

The Founder's ideas of liberty needed to give way to a new liberty for a new time.

This time, liberty was based upon achieving "social order, unity, and development" which was threatened by "dominant economic class" and its attempts to amass wealth. The "social order" that Dewey had in mind could not "be established by an unplanned and external convergence of the actions of separate individuals," i.e., by democracy. This fantasy of popular sovereignty—the core argument on Federalist Number 10—is, to Dewey, the "Achilles heel of liberalism." And, therefore, the new "ends" of the new liberty can be "achieved only by *reversal* of the means to which early liberalism was committed."

Organized social planning, put into effect for the creation of an order in which industry and finance are socially directed in behalf of institutions that provide the material basis for the cultural liberation and growth of individuals, is now the sole method of social action by which liberalism can realize its professed aims. Such planning demands in turn a new conception and logic of freed intelligence as a social force.[84]

This is the sum and substance of the new political order ushered in by Woodrow Wilson: new "institutions" run by experts and possessing unfettered power, charged with providing the "material basis" of liberty. It is unabashedly a new view of law, of liberty, of government, of rights, and even of logic. They were not trying to fulfill the spirit or purpose of the Constitution. They are openly hostile to it.

But one reaction to all this history might be: so what? Even if I have proven that the modern administrative state, with its trust-the-experts mantra, is inconsistent with the foundations of America, what if Wilson, not Madison, was right? Fair enough. Until now, this book has only been concerned with proving that one must choose between Madison and Wilson because they are irreconcilably opposed. Maybe Madison's cachet is not what it used to be, as well as that of all the other "old white guys" that signed America's two existential documents. And so, the remainder of this book will be concerned with

---

[83] J. Dewey, *Liberalism and social action* (New York: Putnam, 1935).
[84] Dewey, *Liberalism and social action.*

debunking Wilson's utopian dream of a country (or, better yet, a world) ruled by efficient and dispassionate experts. We will see that this vision exalts science to a role it cannot possibly fulfill and leads only to Robespierrian tyranny. We will see that Madison was right.

4.

The Excellent Servant and the Terrible Master

For the scientific method can teach us nothing else beyond how facts are related to, and conditioned by, each other. The aspiration toward such objective knowledge belongs to the highest of which man is capable, and you will certainly not suspect me of wishing to belittle the achievements and the heroic efforts of man in this sphere. Yet it is equally clear that knowledge of what is does not open the door directly to what should be. One can have the clearest and most complete knowledge of what is, and yet not be able to deduce from that what should be the goal of our human aspirations. Objective knowledge provides us with powerful instruments for the achievements of certain ends, but the ultimate goal itself and the longing to reach it must come from another source.

—Albert Einstein, *Science & Religion.*

Rationalia

Neil deGrasse Tyson is an accomplished and well-respected astrophysicist and a true American success story. After growing up in the Bronx, Tyson studied physics at Harvard and received his Ph.D. in astrophysics from Columbia University. His work resulted in significant contributions to his scientific field and led him to become the director of the Hayden Planetarium in New York City. In the early 2000's Tyson's success attracted political attention. President George W. Bush appointed him to multiple commissions working on the United States' aerospace industry and the study of the United States' space exploration policy. From there, Tyson became something of a celebrity-scientist in the style of a Carl Sagan or Bill Nye, hosting a variety of scientific television programs, mostly on PBS, and regularly appearing on popular talk shows.

The double whammy of star-and-science power took its toll. In recent years, Tyson at times found himself knee-deep in controversy taking public positions on political hot topics ranging from religion to social justice. In 2016, for example, Tyson waded into politics. He wrote on Twitter that "[e]arth needs a virtual country" that he dubbed "Rationalia" with a "one-line Constitution: All policy shall be based on the weight of evidence."[1]

Tyson doubled down on Rationalia after moderate public backlash suggested that scientific evidence was ill-equipment to make judgments about human values. Tyson penned his political magnum opus in response on Facebook.[2] In his post, he explained that that the types of policies to be settled by "evidence" include "whether a government should help the poor, and if so, in what ways" and "what tax rate should be established, and on what kinds of income." Tyson contrasted the scientific method with the "adversarial" methods of political argument, which, he believed, lead to squabbling and "stall[ed]" policy. And so "in Rationalia," he imagined, the government would "heavily" fund "the sciences that study human behavior (psychology, sociology, neuroscience, anthropology, economics, etc.)" because it is through these sciences that we derive "much of our understanding of how we interact with one another." Therefore, government ought to find its foundation in the day-to-day work of science, allowing "data gathering, careful observations, and experimentation" to influence "practically every aspect of our modern lives." Tyson thus dreamt of a world in which scientific discovery would be "built into the DNA of how the government operations, and how its citizens think."[3]

While Tyson's social media posts about Rationalia were necessarily cursory and basic, the theory of a scientific basis for human values (and, thereby, political action) has been gaining broader support in recent years. It has been fleshed out by some of today's leading academics in New York Times bestselling books, often with the support of some of the most powerful people on the planet. These theorists seek to hew a third path that eschews the two traditional accounts of morality and value: God and relativism. For the right, God provides the basis for metaphysical "goodness," objectively grounding morality, values, purposes, and goals. The left, on the other hand, denies the objectivity of "goodness" and considers values, purposes, and goals as relative to culture or the individual person. Existentialists like Jean Paul Sartre, for example, posit that individuals must determine for themselves their purposes, values, goals, and even their own moral code. His existentialist mantra held that you first exist, and only then can (and should) you define your own essence.

---

[1] K. Artherton, "Neil deGrasse Tyson's Proposed "Rationalia" Government Won't Work," Popular Science, June 29, 2016, Accessed: May 28, 2021, https://www.popsci.com/neil-degrasse-tyson-just-proposed-government-that-doesnt-work/.
[2] K. Artherton, "Neil deGrasse Tyson Doubles Down On Rationalia," Popular Science, Aug. 8, 2016, Accessed: May 28, 2021, https://www.popsci.com/neil-degrasse-tyson-doubles-down-on-rationalia/
[3] Ibid.

Neil deGrasse Tyson and those that have more thoroughly described theories similar to his "Rationalia" deny both of these traditional accounts. Like the right, they believe that the "Good" (morality, values, goals, and purpose) can be objectively discovered. They are not relativists. But like the left, they deny either the existence of, or the need for resort to, something metaphysical like God to provide an objective basis for the Good. This idea is known as ethical naturalism: ethical propositions are made objectively true, not by God or any other metaphysical entity, but by the features of the natural, physical world.

Broadly speaking, ethical naturalism is not a new philosophy. The Eighteenth-Century moral philosophy of utilitarianism is often cast in a naturalist mold, for example, in Jeremy Bentham's *An Introduction to the Principles of Morals and Legislation*.[4] There, Bentham attempted a naturalistic description of morality based on what he called the "felicific calculus," an attempt to quantify and compare the pleasure and/or pain caused by any particular answer to a moral question.

Modern iterations of a natural ethic tend to emphasize the fundamental role that the scientific method, as such, can be brought to bear upon moral inquiries. This idea has been popularized in recent years by the podcaster, professor, and pop philosopher-scientist-atheist, Sam Harris. His aptly titled and bestselling 2010 book *The Moral Landscape: How Science Can Determine Human Values* lays out his case. There, Harris is up-front in acknowledging that "[m]ost people imagine that science cannot pose, much less answer, [moral questions]...No one expects science to tell us how we *ought* to think and behave."[5] Accordingly, he is forced to "confront some ancient disagreements about the status of moral truth: people who draw their worldview from religion generally believe that moral truth exists, but only because God was woven it into the very fabric of reality; while those who lack such faith tend to think that notions of 'good' and 'evil' must be the products of evolutionary pressure and cultural invention." As an ethical naturalist, Harris believes that "both sides in this debate are wrong."[6]

Instead, Harris proposes that "moral truth can be understood in the context of science." The objective basis for morality, in his view, is human "flourishing" or "the well-being of conscious creatures." Therefore, Harris believes that values can be "translate[d] into facts that can be scientifically understood:

---

[4] J. Bentham, *Introduction to the Principles of Morals and Legislation* (Dover Publications, 2012).
[5] S. Harris, *The Moral Landscape: How Science Can Determine Human Values* (New York: Free Press, 2010).
[6] Harris, *The Moral Landscape: How Science Can Determine Human Values*.

regarding positive and negative social emotions, retributive impulses, the effects of specific laws and social institutions on human relationships, the neurophysiology of happiness and suffering, etc."[7] Harris' metaphorical vision is one where morality takes on the form of a landscape: there are many mountainous peaks where a particular moral rule or law will produce objective well-being (some peaks paradisaically higher than others) and many desperate valleys where a different rule impedes or undermines objective well-being (some troughs parade-drenchingly lower than others). On the very first page of his book, Harris discusses the Albanian tradition of vendetta, known as *Kanun*, where a murder victim's family is allowed to kill any of a murder's male relatives in reprisal as an example of a rule that, he believes, we can all agree objectively impairs well-being. Accordingly, Harris proposes that the scientific method can be used to arbitrate the objective elevation of any specific position in the moral landscape: *Kanun*=bad, while bread-for-starving-orphans=good.

For that reason, according to Harris, "morality should be considered an undeveloped branch of science."[8] He touts the scientific research into the use of corporal punishment as a method of disciplining children as an example of science supposedly answering moral questions. "[A]ll the research indicates that corporal punishment is a disastrous practice, leading to more violence and social pathology—and, perversely, to greater support for corporal punishment." Finally, once science has had its say, Harris brings in his version of Tyson's Rationalia: "Clearly, such [scientific] insights could help us improve the quality of human life—and this is where academic debate ends and choices affecting the lives of millions of people begin."[9]

Harris later pushed back on the more Orwellian interpretation of his book's prescription on what to do with moral insights once science achieves them. In an interview with *Salon*'s Katherine Don, he was asked how scientists are supposed to enforce their findings on moral truth, once they are discovered.[10] Harris denied the idea that "scientists in lab coats" would be "at every door." Scientists, he thought, would simply *publish* their findings like they have always done. He replied, "[i]f we learn [how to best teach children], what parent wouldn't want that knowledge? The fear of a 'Brave New World' component to this argument is unfounded."[11]

---

[7] Harris, *The Moral Landscape: How Science Can Determine Human Values.*
[8] Harris, *The Moral Landscape: How Science Can Determine Human Values.*
[9] Harris, *The Moral Landscape: How Science Can Determine Human Values.*
[10] K. Don, "The Moral Landscape: Why science should shape morality," Salon, Oct. 17, 2010, Accessed: May 28, 2021, https://www.salon.com/2010/10/17/sam_harris_interview/
[11] Ibid.

Certainly, scientists are not trained in the use of cudgels and other forms of governmental coercion, but once science has spoken, there may be others who are. If morality is really reducible to a scientific fact—like the earth revolving around the sun—can we really expect the "people-of-science" to tolerate or entertain debate with those that espouse some pre-Copernican moral viewpoint that has been supposedly disproved beyond a scientific shadow of a doubt?

Indeed, after *The Moral Landscape*, the next step for the modern cabal of Moral Scientists has been to puff up science's reputation based on the tremendous advances it has occasioned in its traditional realm of technology and physical discovery. A quote from Harvard psychologist Steven Pinker appears on the cover of my edition of *The Moral Landscape*, asserting that "Harris makes a powerful case for a morality that is based on human flourishing and thoroughly enmeshed with science and rationality." In 2018, it was Pinker's turn. He published his own book, imperatively titled *Enlightenment NOW: The Case for Reason, Science, Humanism, and Progress*. His book was likewise enrobed with praise from many of this ilk: Bill Gates, Nicholas Kristof, David Brooks, and Richard Dawkins.

Enlightenment NOW trumpets a narrow vision of the Age of Enlightenment (the same one that so influenced America's Founding Fathers). Pinker's Enlightenment was a secular one where the scientific method apparently sprung, full-grown, from the head of a secular, naturalistic philosophy. And like Harris, Pinker looks to this version of the Enlightenment to tread a third path between "traditional religious beliefs [that have been] undermined by our best science" and "existential angst."[12] He believes that science can be brought to bear the fundamental question of existence: "Why should I live?" And again echoing Harris, Pinker's answer is that life is worth living because "you have the potential to flourish."

> You can seek explanations of the natural world through science, and insight into the human condition through the arts and humanities. You can make the most of your capacity for pleasure and satisfaction...You can foster the welfare of other sentient beings by enhancing life, health, knowledge, freedom, abundance, safety, beauty, and peace. History shows that when we sympathize with others and apply our ingenuity to improving the human condition, we can make progress in doing so, and you can help to continue that progress.[13]

---

[12] S. Pinker, *Enlightenment Now* (London: Penguin Books, 2019).
[13] Pinker, *Enlightenment Now*.

In other words, according to Pinker, a proper "understanding [of] the human condition" is "informed by modern science" and backed by objective "data."[14]

It is easy to understand the motivation underlying the Moral Science of Tyson, Harris, and Pinker. The star of Science has never been higher as it is today. Science has consistently and continually improved our lives through advancements ranging from technology to medicine. If only there was a way to use its cachet to serve the dual purposes of warding off any commitments to metaphysical and supposedly "superstitious" religion while also making *quasi-religious* claims to objective truth such that others can legitimately be called to lockstep rather than retreating to a timid relativism. But the question remains: has this supposed science of the fundamental questions of life stretched science beyond its useful sphere? Have the Moral Scientists asked science for things that it cannot, even in principle, deliver?

Is's and Ought's

The realm of science envisioned by Tyson, Harris, and Pinker truly knows no bounds. If you want to learn morality, goals, purposes, laws, the very meaning of life itself, look no further than your friendly, neighborhood scientist. If you are confused by some superstitious mumbo-jumbo or charismatic religious zealot, the scientist will suffer no fools and will set you right. Have a dispute with your neighbor? The scientist is the high priest of justice and can settle the matter with an assurance of objective fairness. And yet, all of these questions involve normative values. What should I do with my life? Was it wrong for me to steal from my affluent Scrooge-like neighbor to increase my charitable giving during an economic recession? Should the government increase my neighbor's taxes to accomplish a similarly motivated end?

The premise of the Moral Science is that science can "discover" the answers to these questions by testing the "facts" of the world. In *The Moral Landscape*, Harris asserts that the "divide between facts and values is illusory." Whatever can be known about values, which Harris believes is reducible to well-being, "must at some point translate into facts about brains and their interaction with the world at large."[15] Nonetheless, Harris candidly acknowledges that his position is at odds with the majority of philosophers, including especially one of the most prominent philosophers of the Enlightenment itself, David Hume.

Reacting to similar scientific over-enthusiasm in his *Treatise of Human Nature*, Hume argued that "morality consists not in any relations which are the

---

[14] Pinker, *Enlightenment Now.*
[15] Harris, *The Moral Landscape: How Science Can Determine Human Values.*

objects of science...[and] consists not in any *matter of facts*, which can be discovered by the understanding."¹⁶

> Take any action allow'd to be vicious: Wilful murder, for instance. Examine it in all lights, and see if you can find that matter of fact, or real existence, which you call vice. In which-ever way you take it, you find only certain passions, motives, volitions and thoughts. There is no other matter of fact in the case. The vice entirely escapes you, as long as you consider the object. You never can find it, till you turn your reflexion into your own breast, and find a sentiment of disapprobation, which arises in you, towards this action. Here is a matter of fact; but 'tis the object of feeling, not of reason. It lies in yourself, not in the object... Vice and virtue, therefore, may be compar'd to sounds, colours, heat and cold, which, according to modern philosophy, are not qualities in objects, but perceptions in the mind.¹⁷

Yes, murder involves immense feelings of pain and loss, and in Harris's terminology, a marked decrease in "well-being." But such feelings are like vibrations of the air. Without a metaphysic of morality, the conclusions of outrage, injustice, and immorality are just how we perceive those vibrations. Murder "sounds" bad to us like a toddler banging on piano keys. And charity "sounds" good just like the objective vibrations of air in a well-composed sonata performed by a well-trained musician.

Hume accordingly warned that many philosophers will try to sneak their "oughts" into a series of statements about what "is."

> [Such a] author proceeds for some time in the ordinary way of reasoning...when of a sudden I am surpriz'd to find, that instead of the usual copulations of propositions, *is*, and *is not*, I meet with no proposition that is not connected with an *ought*, or an *ought not*. This change is imperceptible; but is, however, of the last consequence. For as this *ought*, or *ought not*, expresses some new relation or affirmation, 'tis necessary that it shou'd be observ'd and explain'd...But as authors do not commonly use this precaution, I shall presume to recommend it to the readers; and am persuaded, that this small attention wou'd subvert all the vulgar systems of morality, and let us see, that the distinction of

---

[16] D. Hume, D. Norton, M. Norton, and D. Norton, *A treatise of human nature* (Oxford: Clarendon Press, 2014), p. 468.
[17] Hume, Norton, Norton, and Norton, *A treatise of human nature*, p. 468.

vice and virtue is not founded merely on the relations of objects, nor is perceiv'd by reason.[18]

Hume's argument has since become known as the "Is-Ought Problem," which suggests that it is impossible to deduce or infer what one ought to do based solely upon factual propositions of the natural, physical world. Hume's argument has been found persuasive for centuries. Even Harris concedes that Hume's view of the inherent limitation of science is the "received opinion in intellectual circles."[19] Attention to this is-ought problem should, according to Hume, allow us to subvert the "vulgar systems of morality" based on "the relations of objects" and "reason."[20]

Harris, for example, smuggles the "ought" into an "is" with his concept of "well-being." Why is the "well-being" of conscious creatures valuable? Harris simply begs the question, saying "it makes no sense at all to ask whether maximizing well-being is 'good.'"[21] But what can Harris say to the axe-wielding serial murder to prove that his bent toward the destruction of all is simply a factual mistake? If someone doubts that earth's gravitation results in an acceleration of falling objects by 32 feet per second per second, we can bring them into a laboratory and demonstrate this by experiment and measurement. The person who persists in denying the fact after such a demonstration is simply acting in bad faith. But what experiment can demonstrate that the collective well-being of conscious creatures *ought* to be valued by the axe-wielder? All Harris can do is marshal the sentimental schmaltzy stuff of an afternoon television special for the axe-murderer: "Look how beautiful this sunset is…Here's a few Renaissance masterpieces. Aren't they beautiful?... Hey, doesn't that suffering orphan on the street-corner just cry out for justice?" But after we do all that and the axe-wielder still refuses, honestly, to relent, the best Harris can do is label him mentally ill and discard his perspective as invalid.

This problem of proving the value of well-being (indeed, even being itself) is not limited to the hypothetical persuasion of serial killers. Many philosophers have concluded that there is no positive value for human life at all. The Nineteenth Century German philosopher Arthur Schopenhauer argued that one should strive to overcome and deny the will-to-live because life is characterized inexorably by unwarrantable suffering.[22] In this conclusion,

---

[18] Hume, Norton, Norton, and Norton, *A treatise of human nature*, p. 468.
[19] Harris, *The Moral Landscape: How Science Can Determine Human Values*.
[20] Hume, Norton, Norton, and Norton, *A treatise of human nature*, p. 468.
[21] Harris, *The Moral Landscape: How Science Can Determine Human Values*.
[22] A. Schopenhauer, *The world as will and representation* (1818), vol. 1, section 68.

Schopenhauer was heavily influenced by some eastern philosophies and religions. More recently, there has been a philosophical school and moral movement known as "antinatalism" that argues that life is not worth living and that it would have been better for you to have never been born. These philosophers suggest that the superior moral choice in life is to refuse to have children so that the human race will cease to exist. Some Buddhists advance this position.[23] South African philosopher David Benatar explicated this view at length through a western philosophical lens in his book *Better Never to Have Been: the Harm of Coming Into Existence*.[24] What data could Harris show these people, or what experiment could he perform in their presence, to scientifically show that human flourishing, not human extinction, is the objectively correct viewpoint?

There is simply no way to scientifically reconcile the philosophical position of someone like Benatar with ones that affirm life either in religious or non-religious traditions. Either a metaphysical or a relativist commitment to the "good" is required to demonstrate that Benatar has it wrong. The Bible draws the distinction between the value of life and death as based in God:

> This day I call the heavens and the earth as witnesses against you that I have set before you life and death, blessings and curses. Now choose life, so that you and your children may live and that you may love the Lord your God, listen to his voice, and hold fast to him. For the Lord is your life.[25]

And in the non-religious context, there are many existential philosophers, like Jean Paul Sartre, who say one simply *chooses* to side with a life-affirming viewpoint, valuing freedom, adventure, and drinking life into the full. But Sartre would not say that Benatar is objectively wrong to yearn for extinction. Harris would agree with the output of the biblical view that life is good, but he is simply unwilling to accept the religious underpinnings necessary to hold it or to agree with the relativists that it is a matter of choice. He wants to have his cake objectively and eat it non-religiously too.

Science cannot, even in principle, arbitrate between Moses' above-quote and Schopenhauer's existential pessimism. In Hume's parlance, it would be like trying to use science to prove that the way I perceive the color blue is really the same as you do. You cannot derive or infer an "ought" from what "is" in the physical world.

---

[23] H. Gour, *The Spirit of Buddhism* (New Delhi: Cosmo, 1990), pp. 286–288.
[24] D. Benatar, *Better Never to Have Been: the Harm of Coming Into Existence* (Oxford: Oxford University Press, 2006).
[25] Deuteronomy 30:19-20.

The Ambiguity of Well-Being or "Flourishing"

Even if the moral scientists can overcome the is-ought problem, they face an additional obstacle in using the scientific method for matters of human values. Assuming it is an observable fact that we ought to maximize human flourishing, there is still no objective way for a scientist to measure well-being or evaluate competing claims about whether and how much a particular law would bolster it. A cornerstone of the scientific method is falsifiability of hypotheses. For an assertion to be susceptible to the scientific method, it must be one for which we can imagine an experiment or set of conditions under which it could be proven false. It does not necessarily need to be an experiment that is feasible to run today, but it must be objectively conceivable.

Harris, for example, suggests that a person's well-being is a function of their brain chemistry. As such, he thinks it is possible to conceive of experiments to test a moral hypothesis by measuring its impact on a person's brain, or even better, its impact on the all conscious brains in the world. Harris would then compare the "well-being" results of various moral hypotheses and conclude that moral "goods" are defined as those hypotheses that collectively maximize well-being. This sounds well and good in the abstract, but the devil is truly in the details here.

Even if the brain chemistry of every person in the world could be accurately scanned instantaneously and tracked over time, people can reasonably debate whether certain mental states truly reflect "well-being." For example, a committed, intelligent hedonist knows that indulgence in pleasures—good food, alcohol, tobacco, sex, and other luxurious creature comforts—have their price. He knows the "well-being" of his mind after a nice Bordeaux, and he also knows the pains of a hangover and the health risks of overindulgence. But let's say a particular hedonist has done the calculus and believes that well-being is maximized by fairly regular indulgences. Someone that does not believe in hedonism, on the other hand, might set a different value on the hedonist's nice "buzz." Harris's brain experiment cannot break the tie between the hedonist and the non-hedonist. Even with outlandish claims about the insights brain chemistry can afford, can we even imagine an experiment testing such a claim? If some people value the joviality caused by a glass or two of wine, while others simply do not, how can science determine how much joviality is the "optimum"? It is not that the hedonist and the non-hedonist disagree on the quantification of a value. It is not a scientific measurement problem; they just disagree on what should be valued in the first place: the jovial moment or the longevity of life.

Beyond that, there is also the question of the degree to which we value human life itself relative to certain freedoms. Everyone knows that alcohol costs many human lives each and every year. Just think of all the barroom brawls, the car accidents, the liver disease. So let's have Harris test this hypothesis: all other moral rules being equal, well-being will be maximized by total prohibition of alcohol (whether by law or by moral rule—though as discussed, when it is science that makes the determination, what distinction could there be?). How can the scientist weigh the "well-being-values" of nice buzzes and occasions of camaraderie relative to the well-being lost by alcohol's victims? And even beyond that, how does the scientist factor-in the "justice" of certain losses of well-being? Does the well-being of the innocent victim of a DUI accident weigh more than the well-being of an alcoholic who develops cirrhosis of the liver? Does the fact that the alcoholic had a genetic pre-disposition to alcoholism alter the weight of his loss of well-being in the calculus? And then, most obviously of all, how could the scientist even in principle weigh the well-being occasioned by the liberty to drink against the well-being lost by the restriction on freedom? Even if the scientist was honest and had perfect access to the mental states of all involved, the scientist would necessarily be forced to bring personal values to bear on these questions. Even in principle, there are no single, scientifically "correct," answers to these questions.

Like all philosophies that are rooted in utilitarianism, Harris's measurements of brain-chemistry flourishing presuppose that morality is outcome-driven. What matters for Harris in determining the morality of an action are its consequences. But there are many philosophers who believe that it is the action itself (and/or the motivation underlying it) that make it moral or immoral, not its consequences. For example, many would consider charitable donations to be an inherently good moral action. But does your donation become immoral if you were duped into giving to a charity that was actually evil and used your money to fund assassinations? The utilitarian might say yes. The distinction can also be applied to a higher level of rule-based abstraction. Some non-utilitarian philosophers—Immanuel Kant, for example—would argue that lying is always immoral. But maybe Harris's well-being measurement device would determine that well-being is maximized if, all things equal, little white lies are permitted. This measurement of the amount of well-being produced by little white lies would not make one ounce of difference someone like Kant. And maybe you would disagree with Kant. But what data could Harris show Kant to persuade him otherwise? What experiment could he run? Even perfect science could not persuade Kant that lying is sometimes permissible, or even good.

The truth is that so much is built into Harris's little word—"well-being"—that it serves no more insightful purpose than the word around which the whole game of moral philosophy is built: the Good. Harris might as well say that morality is about being Good. But then it sounds like he has not really said much of anything at all (because he hasn't). "Well-Being" is just a black box that he proposes to be filled by scientists because, well, he's a scientist and he is smart and knows what is good and what is bad. In the final analysis, the "data" produced by scientists purporting to measure Well-Being will tell us more about what the scientist thinks is Good than about any objective fact about the world.

The Metaphysics of the Scientific Method

Finally, ethical naturalists like Tyson, Harris and Pinker that seek to ground the big questions of life in the scientific method will encounter one last hurdle. The scientific method is not, itself, natural or scientific.

Albert Einstein observed that the scientific method is really "nothing more than a refinement of everyday thinking."[26] Through science, we use reason to try to comprehend the world and surprisingly find success. "[T]he eternal mystery of the world is its comprehensibility...The fact that [the world] is comprehensible is a miracle."[27] For Einstein, this meant that:

> Physics constitutes a logical system of thought...whose basis cannot be obtained through distillation by any inductive method from the experiences lived through, but which can only be obtained by free invention. The justification (truth content) of the system rests in the proof of usefulness of the resulting theorems on the basis of sense experiences, where the relation of the latter to the former can only be comprehended intuitively.[28]

In other words, Einstein believed that we are justified in believing the assertions of science, not so much because they are "true" in a universal truth-with-a-capital-T sense, but because the assertions of science are useful.

Like Einstein, Harris also appears to accept that the scientific method itself might not be a scientifically measurable fact, but a value, based on what he perceives as its usefulness. He concedes that "the very idea of 'objective' knowledge (i.e., knowledge acquired through honest observation and reasoning) has values built into it, as every effort we make to discuss facts depends upon principles that we must value (e.g., logical consistency, reliance on evidence, parsimony, etc.)."[29] But if scientific methodologies represent a value, then their results must be value-laden and open to criticism by those with different values. A marginalized critic of science might say, "sure, these

---

[26] A. Einstein, "Physics and reality," *Journal of the Franklin Institute* 221, 3 (1936): pp.349-382.
[27] Einstein, "Physics and reality," pp.349-382.
[28] Einstein, "Physics and reality," pp.349-382.
[29] Harris, *The Moral Landscape: How Science Can Determine Human Values*.

scientific assertions are useful to privileged people like Einstein and Harris, but they have never been useful to me, never put a scrap of money in my pocket. So if they believe that science should only be accepted because it is 'useful,' why can't I reject it as useless to me?"

Religious scientists, on the other hand, can explain Einstein's "miracle" of the comprehensibility of the world with reference to God and the objective, universal, capital-T Truth of reason, logic, and the scientific method. The theist claims that God laid a rational metaphysical superstructure upon the base materials of the world. Scientists are tapping into that superstructure when they reason and honestly employ the scientific method. Thus, the success encountered by scientists is not at all surprising in this account. Plato called this superstructure the realm of the Forms. Christians call it the λόγος, the Word of God, the divine pattern etched into the universe with which we can commune. But both Platonists and Christians saw that the patterns by which the world is comprehended cannot be any naturally-occurring thing in the world, objectively measurable by science. They are, by definition, metaphysical—not true just because they are useful to us, but because they are objectively True.

The original scientists of the Scientific Revolution, who presaged the Enlightenment so valued by Harris and Pinker, understood the metaphysical implications of a scientific method. Sir Francis Bacon was the foundational figure in the new empirical science. In the Sixteenth and Seventeenth Centuries, he was a preeminent philosopher, statesman, and scientist in England, rising to the offices of Attorney General and Lord Chancellor. Voltaire quite correctly called him the "father" of the scientific method. Bacon set up his empirical scientific method in purposeful and direct contradistinction to Aristotle's more philosophically-oriented physics. He titled his 1620 book on the new science, *Novum Organum*, in reference to Aristotle's classic *Organon*, and there laid out the logical system by which the new science would be done. Bacon's goal in laying out an *empirical* scientific method was to dampen the flaws of fallen human reason in order to see God's logic stamped into the world:

> For I am building in the human understanding a true model of the world, such as it is in fact, not such as a man's own reason would have it to be; a thing which cannot be done without a very diligent dissection and anatomy of the world. But I say that those foolish and apish images of worlds which the fancies of men have created in philosophical systems must be utterly scattered to the winds. Be it known then how vast a difference there is (as I said above) between the idols of the human mind and the ideas of the divine. The former are nothing more than arbitrary abstractions; the latter are the Creator's own stamp upon creation, impressed and defined in matter by true and exquisite lines. Truth, therefore, and utility are here the very same things; and works

themselves are of greater value as pledges of truth than as contributing to the comforts of life.[30]

We see in the *Novum Organum* the key difference between Bacon and those later scientists who found themselves embarrassed by the metaphysics of the very scientific method they operated under. Einstein and Harris believe that science is good only because it is useful. But, for Bacon, science is useful to the extent it is True, i.e., to the extent its methods uncover the "ideas of the divine" and "the Creator's own stamp upon creation." The utility of science in improving the "comforts of life" is merely a happy side-effect of the search for Truth itself.

Bacon conceptualized the true scientist as a lover, not a utilitarian engineer. The "inquiry of truth" was the "love-making or wooing" of truth. [31] The "knowledge of truth" was its "presence." And, most importantly, the "belief of truth" was the "enjoying of it" and represented the "sovereign good of human nature."[32] Bacon famously described those that care only for utility as "jesting Pilates," alluding to Pilate's famous "What is Truth?" retort to Jesus' claim that he came to the world to testify to the Truth.[33]

Neither was Bacon an outlier in valuing truth over utility in his view of the scientific method. On the Continent, Galileo put it this way: "When God produces the world, he produces a thoroughly mathematical structure that obeys the laws of number, geometrical figure and quantitative function. Nature is an embodied mathematical system."[34] Indeed, some of the scientific pioneers of the recent century preserved those same ideas. Warner Heisenberg, for example, was a foundational figure in the discovery of Quantum Mechanics—lending his name to the so-called Heisenberg Uncertainty Principle. For him, the metaphysical superstructure of nature was more "real" than human notions of "matter."

> [M]odern physics has definitely decided for Plato. For the smallest units of matter are not physical objects in the ordinary sense of the word: they are forms, structures, or—in Plato's sense—Ideas, which can be unambiguously spoken of only in the language of mathematics.[35]

---

[30] F. Bacon, *The Novum Organum of Sir Francis Bacon, Baron of Verulam, Viscount St. Albans, Epitomiz'd for a Clearer Understanding of his Natural History* (1620), at CXXIV.
[31] M. Scott, *The Essays of Francis Bacon* (New York: Charles Scribner's Sons, 1908), at *Of Truth*, p.5.
[32] Ibid.
[33] Ibid.
[34] J. Collins, *A History of Modern European Philosophy* (Bruce Publishing, 1965), p. 81.
[35] K. Wilber, *Quantum Questions* (Shambala, 1984), p.185.

Contrary to the original scientists, a naturalist like Harris or Pinker denies the capital-T truth of metaphysical entities and structures. In Einstein's terms, they are "free" human inventions whose truth content (to the extent it can be called that) exists solely in relation to its "usefulness." This explanation of truth-as-utility is needed by Harris in order to maintain an ethical naturalism; if something metaphysical needed to exist to make the scientific method true, then Harris could not be a naturalist.

But cracks can start to form in that word "useful," leading now to a deluge. Even Harris admits that whether a concept is useful is a "value." Maybe what Harris finds useful, another person will find to be useless, or even worse, a tool of oppression. And when that happens, a purely naturalistic science can have nothing further to say, having disowned the very metaphysics that produced it.

This criticism of a supposedly objective, naturalistic system of value based on the scientific method is by no means a mere hypothetical. In the 1960s and 70s, post-modern philosophers attacked science on the horns of the dilemma that science itself is either unjustified by circular logic or grounded in antiquated metaphysics involving God. Jean-François Lyotard was one of the most preeminent postmodern philosophers in France during this time period. In his 1979 book *The Postmodern Condition: A Report on Knowledge*, he applied the postmodern critique to the scientific metanarrative.

Lyotard observed that the West has "mount[ed] a narrative of scientific knowledge" based ultimately in "sociopolitical authority" where "those who refuse to accept the rules, out of weakness or crudeness, are excluded."[36]

> Scientific knowledge cannot know and make known that it is the true knowledge without resorting to the other, narrative, kind of knowledge, which from its point of view is no knowledge at all. Without such recourse [to non-scientific knowledge] it would be in a position of presupposing its own validity and would be stooping to what it condemns: begging the question, proceeding on prejudice. But does it not fall into the same trap by using narrative as its authority.[37]

Lyotard explained that the original scientists—like Aristotle—were upfront that the basis for science was not scientific, but metaphysical. Aristotle was honest in "separating the rules to which statements declared scientific must conform [in the *Organon*]...from the search for their legitimacy in a discourse on Being

---

[36] J. Lyotard and G. Bennington, *The Postmodern Condition: A Report on Knowledge* (Minneapolis: Univ. of Minnesota Press, 2010).
[37] Lyotard and Bennington, *The Postmodern Condition: A Report on Knowledge*.

[that he called] the *Metaphysics*."[38] As discussed, Bacon did so as well, founding empirical science upon the rock of the λόγος of a divine reality.

Lyotard saw that modern scientists, however, had sought to kick the ladder out from under them. In conceptualizing "scientific knowledge" as purely "pragmatic," they had tried to "leave[] behind the metaphysical search for a first proof or transcendental authority as a response to the question: How do you prove the proof?"[39]

> [S]ince "reality" is what provides the evidence used as proof in scientific argumentation, and also provides prescriptions and promises of a juridical, ethical, and political nature with results, one can master all these games by mastering "reality."…This is how legitimation by power takes shape. Power…legitimates science and the law on the basis of their efficiency, and legitimates this efficiency on the basis of science and law. It is self-legitimating.[40]

In modern science, "there is no other proof that the rules [of science] are good than the consensus extended to them by the experts." Thus, without a metaphysic, it is ultimately just the say-so of the participants in the scientific community (who named themselves experts in a program that they invented) that could enforce or legitimize the results of science. "It is natural in a narrative problematic for such a question to solicit the name of a hero as its response: *Who* has the right to decide for society? Who is the subject whose prescriptions are norms for those they obligate?"[41]

If scientific claims are inherently value-laden, who are Einstein and Harris to decide their "utility"? The response from the scientist is basically what appears in Pinker's *Enlightenment NOW*: here are the "heroes" who legitimize scientific knowledge. Perhaps its Einstein or Jonas Salk. Look at all the benefits of the Enlightenment. We have electricity, vaccines, smart phones, cheap automobiles and air-travel, etc. etc. And all of that may well be persuasive as a narrative. But the point is that someone like Lyotard does not *need* to be persuaded as he must be of objective truths like 2+2=4. Lyotard could point to those that lost out in modernization as proof that he need not accept the objective utility the scientist puts on offer. Thus postmodernists like Lyotard were dexterous and skilled in "delegitimizing" such narratives by pointing to the left behind, the marginalized, and the oppressed by the narrative of *Enlightenment NOW*.

---

[38] Lyotard and G. Bennington, *The Postmodern Condition: A Report on Knowledge*.
[39] Lyotard and G. Bennington, *The Postmodern Condition: A Report on Knowledge*.
[40] Lyotard and G. Bennington, *The Postmodern Condition: A Report on Knowledge*.
[41] Lyotard and G. Bennington, *The Postmodern Condition: A Report on Knowledge*.

Lyotard, for example, suggests that scientists may be in the pocket of a country's wealthy interests. He points out that the day-to-day practice of science is most often performed within a capitalist context. "[S]cience becomes a force of production...a moment in the circulation of capital...The games of scientific language become the games of the rich, in which whoever is wealthiest has the best chance of being right. An equation between wealth, efficiency, and truth is thus established."[42] The "needs of the most underprivileged" will be ignored by science, according to Lyotard. And ultimately, the powerful who wield the cudgel of science will be allowed to undemocratically define a society's desires and values—essentially redefining what it means to be human in the process:

> In this sense, the system [of science] seems to be a vanguard machine dragging humanity after it, dehumanizing it in order to rehumanize it at a different level of normative capacity. The technocrats declare that they cannot trust what society designates as its needs; they 'know' that society cannot know its own needs since they are not variable independent of the new technologies. Such is the arrogance of the decision makers—and their blindness....The decision makers' arrogance, which in principle has not equivalent in the sciences, consists in the exercise of terror. It says: "Adapt your aspirations to our ends—or else."[43]

Lyotard and the post-modernists were drawing on a significant body of Twentieth-Century philosophy in their concern over the unrestrained power of technocrats. The so-called Frankfurt School that arose after World War II anticipated these worries in the decades before the rise of post-modernist philosophy. Before Lyotard leveled his aim at science, Max Horkheimer and Theodor Adorno had criticized the Enlightenment narrative in their 1947 *Dialectic of Enlightenment*. In their view, the "impartiality of scientific language" had sought cover under "neutral sign[s]" to thereby deprive the "powerless of the strength to make itself heard...Such neutrality is more metaphysical than metaphysics."[44] Horkheimer and Adorno argued that the Enlightenment had "devoured...universal concepts, and left nothing of metaphysics behind except the abstract fear of the collective from which it had sprung." Enlightenment had thus turned "itself into an outright deception of the masses" in the name of "power itself."[45] Similar themes were developed as

---

[42] Lyotard and G. Bennington, *The Postmodern Condition: A Report on Knowledge*.
[43] Lyotard and G. Bennington, *The Postmodern Condition: A Report on Knowledge*.
[44] T. Adorno and M. Horkheimer, *Dialectic of Enlightenment* (New York: Herder and Herder, 1947).
[45] Ibid.

to the Enlightenment cornerstones of reason and logic in Horkheimer's *Eclipse of Reason.*

These same criticisms of science, reason, logic, and the Enlightenment continue to the present, often recast within the new politics of the day. In 2008, Tukufu Zuberi and Eduardo Bonilla-Silva published a collection of essays under the title "*White Logic, White Methods: Racism and Methodology.*" In their introduction to these materials, the authors explain their position that mathematics and scientific tools like statistics "developed in relation to European contact, colonization, trade, and domination" of other peoples, and their use in social science "developed as part of the eugenics movement."[46] There is no metaphysical justification, in their view, showing that statistics ought to be "used by sociologist, economist, or political scientist." Echoing Lyotard, they observed that statistical methods are used simply "because of a consensus among the practitioners," *i.e.*, it is a matter of power, not objective logic. And since these authors view power through the lens of race, they view science as the "(White) methodology" behind which the powerful remain "hidden." Under their view, in science, "whiteness becomes normative" and "works like God."[47]

Similar articles and books in the most recent decade abound. A few titles will serve to underscore the point: *The Racism in Science's DNA: The Bad Logic of Aristotle and James Watson* by Velvet Yates;[48] *Making Black Women Scientists under White Empiricism: The Racialization of Epistemology in Physics* by Chanda Prescod-Weinstein;[49] and *Snow Brown and the Seven Detergents: A Metanarrative on Science and the Scientific Method* by Banu Subramaniam.[50]

Science would not have opened itself to the criticisms of the Frankfurt School, the postmodernists, or the recent scholarship of postcolonial and critical race theory if it had not overextended and kicked the metaphysical ladder out from under itself. Once science becomes normative, rather than objective and universal, everything is up for grabs. The Frankfurt School can say that science is an instrument of the capitalist. The postmodernist can say it is the cudgel of the powerful. And the critical race theorist can say it is a tool of racist

---

[46] T. Zuberi and E. Bonilla-Silva, *White Logic, White Methods: Racism and Methodology* (Lanham: Rowman & Littlefield, 2008).
[47] Ibid.
[48] V. Yates, "The Racism in Science's DNA," Medium, 2019, Accessed May 28, 2021, https://eidolon.pub/the-racism-in-sciences-dna-e82bb7638c35
[49] C. Prescod-Weinstein, "Making Black Women Scientists under White Empiricism: The Racialization of Epistemology in Physics," *Signs: Journal of Women in Culture and Society* 45, 2 (2020): pp.421-447.
[50] B. Subramaniam, "Snow Brown and the Seven Detergents: A Metanarrative on Science and the Scientific Method," *Women's Studies Quarterly* 28, 1/2 (2000).

oppression. If Tyson, Harris, and Pinker insist on a *normative* use of science, these criticisms are completely understandable, indeed, inevitable. It is only a humble science that confines itself to Truth, objective facts, and universal reason that may lay claim to transcending the criticism of a Horkheimer, a Lyotard, or a Zuberi. If science is just that which is "useful" or that around which consensus forms, the question will always be: useful to whom and consented to by whom? But if science restrains itself to be satisfied with only that which is True, regardless of its utility, then Bacon's retort is correct. Those that would attack such a science would be "jesting Pilates," left to only undertake the herculean task of proving that there is no such thing as "Truth."

Conclusion: Science *and* Liberty

To invert Marc Antony's funeral oration from *Julius Cesar*: I came to praise Science, not to bury it. But in order to praise Science, it cannot cross the Rubicon and become Cesar. Tyson's *Rationalia* would exalt science to a normative role that it cannot possibly fulfill. Science is great at determining what *is*, but not what *ought to be*.

A wall must be maintained between the realm of Science and the realm of values and morality. Stephen Jay Gould called this the "nonoverlapping magisteria" of science and religion.[51] Science overreached as religion began to decline and as philosophers became skeptical of metaphysical realities beyond the natural world. This blurred the line between the True and the Useful. Bacon's science sought the Truth, which had a happy side effect in enabling things that people found to be useful. For example, quantum mechanics discovered certain truths about light which led engineers to design lasers. But metaphysical/religious skepticism led to a definition of Truth as whatever proves useful in explaining the world (recall Einstein's *Physics and Reality*). For example, a scientist like Harris might not really care whether science's hypotheses of "dark matter" are true in an absolute sense; he just cares that the hypothesis fits and adequately explains data to a useful end.

From there, it is no large leap from that which is useful in explaining data about the physical world to that which is useful in ordering our lives, for example, to maximize "well-being." In bridging that gap, science exceeded its inherent limitations in venturing from the physical to the moral. And it is there that Science must yield to Liberty. As the United States Supreme Court put it: "At the heart of liberty is the right to define one's own concept of existence, of

---

[51] S. J. Gould, "Nonoverlapping Magisteria," *Natural History* 106 (March 1997): pp. 16-22.

meaning, of the universe, and of the mystery of human life."[52] On one side of the political spectrum, these concepts find their fulfillment objectively in God and an objective Truth. On the other side, they are fulfilled subjectively through a personal, existential definition of values.

Liberty would be destroyed if a government could establish a religion and justify law based on religion. In the United States, this is precluded by the First Amendment. But Science functionally serves the same purpose as a religion if a law could be justified on the basis that "science says" we ought to require X or prohibit Y. The violence to the principles of liberty are the same, and the Founders did not anticipate that science would elevate itself to the status of religion.

Therefore, as expected, the criticisms of Science began to flood in. Initially, the most stinging of these criticisms came from a secular context on the left. The Frankfurt School, the postmodernists, and their more recent successors also believed that there is no such thing as truth with a capital-T. Science was only a system that cannot, even in principle, provide its own theoretical justification. If science was only self-justified by its usefulness, then it is fair to consider who, in particular, it was more useful to. And when the answer to that question is believed to be the powerful and the rich, it is fair to want to expose that situation and seek to alter science itself to make it more "equitable." Without a viable distinction between facts and values, the postmodernists could even use their critique to push science out of its home-realm of facts and nature.

It does not have to be this way if only we were to accept the distinction between facts and values, between Science and Liberty. The scientific method is True because it conforms to the way in which the physical world is actually structured. Scientific results, if honest and accurate, are not just useful. They are not justified only by consensus of scientists or only by those humans with the power to bless them. They are justified because they are objectively, universally true. Of course, various scientific theories can be better or worse approximations of the truth. Newton's equations of gravitation were certainly closer to the truth than Aristotle's. And Einstein refined Newton's equations yet further, though even they are unlikely to be the last word. The key, though, is that under this metaphysical system of science, usefulness and power have no say. All that matters is Truth about the physical world.

Just as science is justified by the laws of the natural world, liberty is justified by the natural rights of people. Government must serve liberty by protecting

---

[52] *Planned Parenthood of Southeaster Pa. v. Casey*, 505 U.S. 833 (1992). This book ventures no opinion on the application of that principle in the context of that particular case.

and enhancing those natural rights. Government cannot be the handmaiden of science as Tyson envisioned for Rationalia. Such a government necessarily destroys the People's freedom of values and their power to instantiate such values (within the confines of liberty) through democratic law. Since Science cannot derive or infer an "ought" from an "is," its normative prescriptions are necessarily the preferences of the scientist and not of the scientific method.

Science is, and always should remain, the handmaiden of the People for the increase of objectively true knowledge. What the People choose to do with such knowledge is their province, as a matter of Liberty.

5.
The False Idols

O Sacred, Wise, and Wisdom-giving Plant,
Mother of Science, Now I feel thy Power
Within me clear, not only to discern
Things in their Causes, but to trace the ways
Of highest Agents, deemed however wise.
Queen of this Universe, do not believe
Those rigid threats of Death; ye shall not Die:
How should ye? by the Fruit? it gives you Life
To Knowledge...

      Satan to Eve—John Milton, *Paradise Lost*, Bk. IX:679-687

 True and honest science is very difficult to do. We are well-adapted to survival in the environment, the avoidance of predators, and socialization with other humans. We are not dispassionate computers. We have emotions and peculiarly subjective ways of looking at the world. How difficult and unintuitive it is for a child to understand that objects in the world are not really colored—that is only how light interacts with or bounces off the object and then is interpreted by our brains. Consider how long it took for people to get the distinction between gravitational force and wind-resistance. Who could forget those elementary-school science demonstrations of a feather and rock being dropped in a vacuum?

 Even well-educated adults can find it difficult to understand or conceptualize advanced scientific and mathematical concepts: quantum wavefunctions, space-time and relativity, a four-dimensional hypercube. It recalls the episode of *The Simpsons* where two-dimensional cartoon Homer stumbles through a wall of his house into an abstract, computer-generated three-dimensional landscape. On the other side of the wall, Homer's wife, Marge, consults Professor Frink about what is going on. Frink tells her: "Well, that should be obvious to even the most dimwitted individual who holds an advanced degree in hyperbolic topology—that Homer Simpson has stumbled into...the Third Dimension." Marge gasps, eliciting the cliché response: "Are you saying what I think you're saying, Professor Frink?" Frink dryly replies, "I doubt it, ma'am. It's highly complicated."[1]

We have seen in chapters two and three that rule by science is at odds with the foundational principles of the Declaration of Independence and the Constitution. We saw in chapter four that normative science is also a contradiction in terms. Scientists have no superior claim to rule because their value-laden judgments are necessarily just as human as anyone else's. The scientist can certainly provide insights into scientific facts. However, in a democracy, the People as a whole, comprising scientists on an equal footing, must be allowed the liberty of filtering scientific facts about the world through their own lenses of values.

But even further than the facts-values distinction, do scientists have a monopoly on the ability to either accept or criticize supposedly scientific findings? Scientists are human too, and prone to error. It is obvious that sometimes a scientific error may be more easily discovered by another expert or the scientific community at large than a layman. But other times, the scientific community might be disposed to fail to recognize its own errors. The laity has an equal capacity for reason and common sense, which they can sometimes bring to bear to challenge scientific fallacies that may have blinded the experts. Thus, not only do the People serve as the ultimate arbiters of the Good, but they also have a role to play in working with the scientific community to discover the truth about the natural world.

There are many embarrassing facts that people in insular communities seem to be blinded to. It is often the role of the outsider to point out the truth. Hans Christian Andersen wisely has an outsider to the world of political power—a child—blurt out the truth in his story, *The Emperor's New Clothes*. Similarly, in science, Albert Einstein was working as an examiner at the Swiss Patent Office in 1905 when he published four scientifically earth-shattering papers on the photoelectric effect of light, proof of the particle properties of matter, his theory of special relativity, and mass-energy equivalence. The first person who saw microorganisms through a microscope, Antonie Philips van Leeuwenhoek, was a lay clothier and developed an interest in lenses to help him examine textiles. Neither was enmeshed in the avantgarde of science when they made these discoveries. These are just a few examples where outsiders have more easily added value to an insider-group discussion that had become stale and blind to error.

There are many explanations for why the scientific community can fail. They are all a function of the fact that, like all communities, the scientific community

---

[1] *Simpsons, Treehouse of Horror VI.* [video] Directed by B. Anderson and D. Mirkin (1995).

is made up of people who are human and therefore fallible. Like all of us, scientists can become set in their ways, prideful, and blinded to what might be obvious to others. Like anyone, scientists can be overly reliant on, or deferential to, authority and fearful of going against the grain. They can be subject to institutional pressures and biases not felt by the ordinary person. They can be influenced by money or power. And they can have their own biases, motivations, and agendas. Scientific knowledge, like liberty and moral knowledge, is nurtured and increased by the free and open exchange of ideas without prejudice or overzealous exclusions on the basis of "expertise" or the mantra to fall in line and "follow the science."

Bacon's Idols

Sir Francis Bacon was the founding philosophical expositor of modern empirical science. He did not worship or idolize received scientific knowledge or the "consensus" of scientists or other groups of the "elite." As a forerunner of Enlightenment thinking, he employed, and even indulged, a healthy skepticism toward human endeavors. Imagining himself as a father figure to a rhetorical son in one of his early works on science, Bacon wrote: "The fact is, my son, that the human mind in studying nature becomes big under the impact of things and brings forth a teeming brood of errors."[2]

Science must be objective and dispassionate, but people are fallen, limited creatures that always bring a lot of baggage with them to the scientific enterprise. Bacon observed that "all the approaches and entrances to men's minds are beset and blocked by the most obscure idols—idols deeply implanted and, as it were, burned in." An aspiring scientist will bring these idols "to the interpretation of nature, snatching at any facts which fit in with his preconceptions and forcing everything else into harmony with them." He doubted that "any clean and polished surface remains in the mirror of the mind on which the genuine natural light of things can fall." Therefore, since men are not naturally disposed to conducting science properly, "a new method must be found for quiet entry into minds so choked and overgrown." This method must be "mild and afford no occasion of error." "It must have in it an inherent power of winning support and a vital principle which will stand up against the ravages of time."[3] This is what ultimately became the scientific method.

---

[2] B. Farrington and F. Bacon, *The Philosophy of Francis Bacon* (Chicago: University of Chicago Press, 1964), at *The Masculine Birth of Time* (1603), pp. 60-72.

[3] Farrington and Bacon, *The Philosophy of Francis Bacon*, at *The Masculine Birth of Time* (1603), pp. 60-72.

In 1620, Bacon completed his classic treatise on the scientific method with the publication of *Novum Organum* in which he proceeded by numbered aphorisms written in Latin. Building on his prior work, he argued that the "cause and root of nearly all evils in the sciences" is that people "falsely admire and extol the powers of the human mind" and "neglect to seek for its true help[ers]."[4] Compared to nature, an individual human mind is feeble and easily deceived. "The subtlety of nature is greater many times over than the subtlety of the senses and understanding."[5]

The first helper for the mind is logic and reason. However, while logic is sound, philosophers and scientists had, to that point, employed logic within the framework of human language. "The syllogism consists of propositions, propositions consist of words, words are symbols of notions. Therefore if the notions themselves (which is the root of the matter) are confused and overhastily abstracted from the facts, there can be no firmness in the superstructure" of the logic. The scientist must allow reason and logic to be applied as directly as possible to nature itself. The single best way for getting facts into minds without warping is to "lead men to the particulars themselves, and their series and order."[6] And to do this, the scientist must be wary of false idols:

> The idols and false notions which are now in possession of the human understanding, and have taken deep root therein, not only so beset men's minds that truth can hardly find entrance, but even after entrance is obtained, they will again in the very instauration of the sciences meet and trouble us, unless men being forewarned of the danger fortify themselves as far as may be against their assaults.[7]

Bacon identified "four classes of Idols which beset men's minds" on the pathway to true science. The first is the Idols of the Tribe, which are those aspects of human nature itself that impair the conduct of science. Human understanding is "like a false mirror which received light [from nature] irregularly, then distorts and discolors the nature of things by mingling its own nature with it."[8] Thus, the true scientist must work "according to the measure of the universe," not the measure of humans.

---

[4] F. Bacon, *The Novum Organum of Sir Francis Bacon, Baron of Verulam, Viscount St. Albans, Epitomiz'd for a Clearer Understanding of his Natural History* (1620).
[5] Bacon, *The Novum Organum*.
[6] Bacon, *The Novum Organum*.
[7] Bacon, *The Novum Organum*.
[8] Bacon, *The Novum Organum*.

The second is the Idols of the Cave, which are the idols possessed by the individual aspiring scientist. Everyone "has a cave or den of his own, which refracts and discolors the light of nature." This can be a result of his own peculiar nature, his upbringing, his experiences, or the "authority of those whom he esteems and admires."[9]

The third is the Idols of the Market Place, which arise from "the commerce and consort of men." The aspiring scientist is only able to communicate with other others through language. However, just as the use of words in logical syllogisms can lead to trouble, words can "force and overrule the understanding, and throw all into confusion, and lead men away into numberless empty controversies and idle fancies."[10]

Last are the Idols of the Theater, which are biases and distortions caused by "various dogmas or philosophies." Bacon view these as theatrical because they are like "stage plays, representing worlds of their own creation after an unreal and scenic fashion." Many axioms of science, he thought, had "by tradition, credulity, and negligence" come to be nothing but theater, rather than real insights into the real world.[11]

Bacon knew that upending these idols was not a one-and-done deal for his fledgling science. They were bound to "meet and trouble us" again and again, and we must keep vigilant and everlasting watch to prevent their incursions back into the scientific community.

Have scientists faithfully kept this watch as the decades and centuries of experience with the scientific method have worn on after Bacon? Recent years have seen the application of the scientific method well beyond the natural world in the so-called social sciences. Do the idols have a nearly insurmountable hold on those areas of human inquiry?

What follows is a discussion of some modern scientific crises viewed through the lens of Bacon's four classes of false scientific idols. This will not be anti-science; the scientific method is and continues to be the best tool people have devised to investigate the natural world. The following is only anti-idol, against a warping of the scientific method for non-scientific ends. And the question I raise is whether the People, through critical analysis and healthy skepticism, have a legitimate role to play in the evaluation of what is presented as received scientific truth.

---

[9] Bacon, *The Novum Organum*.
[10] Bacon, *The Novum Organum*.
[11] Bacon, *The Novum Organum*.

The Idols of the Tribe

The Idols of the Tribe are those commonly-held and natural human weaknesses or propensities that tend to resist a truly dispassionate application of scientific method. For example, Bacon warned that it is in the nature of "human understanding" that "when it has once adopted an opinion (either as being the received opinion or as being agreeable to itself), [it] draws all things else to support and agree with it."[12] Even when a person is confronted with growing evidence against a preconceived notion or belief, the mind "either neglects and despises [such evidence], or else by some distinction sets [it] aside and rejects [it] in order that by this great and pernicious predetermination the authority of its former conclusions may remain inviolate." In this way, a person's "first conclusion colours and brings into conformity with itself all that come after." In other words, even scientists are human, and they can tend to blind themselves to data that conflicts with a previously held belief. Human reasoning is "no dry light, but receives an infusion from the will and affections."[13]

Beyond this, it is in people's nature when encountering data or empirical evidence "to suppose the existence of more order and regularity in the world than it finds." People are quick to draw conclusions about causation. It is almost a cliché to hear the retort, "well, correlation is not causation." Sometimes this insight can be misapplied and overused dismissively. But far too often, the correlation-causation mistake is made, even by scientists. This is because, as Bacon saw, "human understanding is moved by those things most which strike and enter the mind simultaneously and suddenly, and so fill the imagination."[14]

Ironically, we all know the way to defeat the Idols of the Tribe. It is integrity. Absolute, ruthless, self-effacing integrity. The best way to discover our own lapses of reason is to crowd-source what we have blinded ourselves to. The problem is that practicing this type of integrity is incredibly difficult because it runs up against that most fundamental of human vices: pride. The type of integrity that is needed to defeat the Idols of the Tribe is a sort of naked vulnerability. It means we need to arm our potential critics with the best possible weapons that they can turn on, and use against, us. And very few people, scientists and myself included, are able to consistently do this. In other words, at the beating heart of true science is a desire not to prove your hypothesis, but to disprove it, accompanied by pleasant surprise when you fail.

---

[12] Bacon, *The Novum Organum*, at XLVI.
[13] Bacon, *The Novum Organum*, at XLVI.
[14] Bacon, *The Novum Organum*, at XLVI.

The False Idols

The renowned Twentieth Century physicist Richard Feynman coined a helpful term for when science yields to the Idols of the Tribe. He called it Cargo Cult Science, which he discussed during his 1974 commencement address to that year's class of graduates at the California Institute of Technology.[15] Cargo cults are quasi-religious rituals that have appeared in some societies of the developing world after they have had interactions with more developed countries. These societies sometimes formed religious cults whose goal was to obtain the materials goods (the "cargo") of the developed countries by mimicking what they perceived as their practices (the "cult"). Feynman gave this example:

> During the war [some South Sea Islanders] saw airplanes land with lots of good materials, and they want the same thing to happen now. So they've arranged to make things like runways, to put fires along the sides of the runways, to make a wooden hut for a man to sit in, with two wooden pieces on his head like headphones and bars of bamboo sticking out like antennas—he's the controller—and they wait for the airplanes to land. They're doing everything right. The form is perfect. It looks exactly the way it looked before. But it doesn't work. No airplanes land. So I call these things Cargo Cult Science, because they follow all the apparent precepts and forms of scientific investigation, but they're missing something essential, because the planes don't land.[16]

To Feynman, these cargo cults were not just some interesting anecdote. They revealed something about our universally shared human nature that continues to exist today in the way modern science is being done. And it served as a reminder of how the scientific method had found a way to fight against human nature to achieve the results people were looking for.

The cure to Cargo Cult Science, according to Feynman, was "scientific integrity," "utter honesty—a kind of leaning over backwards." Applied to science, this means that "if you're doing an experiment, you should report everything that you think might make it invalid—not only what you think is right about it." This includes trying to list other causes that might explain the results or even things that the scientist consciously tried to control for, so that others can see for themselves whether they think the control was adequate. Scientists must resist the impulses of pride and affirmatively give the scientific community "[d]etails that could throw doubt on [their] interpretation" of the

---

[15] R. Feynman and E. Hutchings, *Surely you're joking, Mr. Feynman!* (New York: W. W. Norton, 1997), p. 338.
[16] Feynman and Hutchings, *Surely you're joking, Mr. Feynman!*, p. 338.

experiment and recite "all the facts that disagree" with the scientists' proposed theory. Finally, such integrity will lead the scientist to try to find whether a theory truly has predictive power, rather than just fitting with the facts from which the scientist designed the theory. In other words, the scientist will allow, not only other people, but also the natural world itself an opportunity to test the theory and prove it wrong. "The first principle is that you must not fool yourself [because] you are the easiest person to fool."[17]

Feynman also explained that this kind of integrity also applied in science's relationship with the People. There might be a tendency for scientists to portray themselves as infallible to the public. It can be too easy to dismiss the challenges or criticisms of lay person on the basis of a lack of expertise, or even intelligence. That human tendency might be fought against and corrected. When dealing with "laymen," a scientist should engage in "a specific, extra type of integrity that is not lying, but bending over backwards to show how you're maybe wrong" to avoid "fool[ing] the layman." "If you're representing yourself as a scientist, then you should explain to the layman what you're doing—and if they don't want to support you under those circumstances, then that's their decision."[18]

Despite the scientific ideals found in the value of skepticism and humility as espoused by Feynman, we often find precisely the opposite in the real world, especially as it relates to the proposed role of the People relative to scientific claims. Naomi Oreskes, a historian of science, argues that firm "trust" in science is imperative and justified wherever the scientific community forms a "consensus." In her book, *Why Trust Science?*, she warns that "[i]f trust in experts were to come to a halt, society would come to a halt, too. Scientists are our experts in studying the natural world and sorting out complex issues that arise in it."[19] Oreskes rejects the idea that science is fundamentally characterized by falsifiability. Instead, "[a] claim that has survived critical scrutiny becomes established *fact*, and collectively the body of established facts constitute scientific *knowledge*."[20] Therefore, scientists "can do things we can't. And just as we wouldn't go to a plumber to fix our teeth…[i]f we need scientific information, we should go to the scientists who had dedicated their lives to learning about the matters at stake. On scientific matters, we should trust science."[21]

---

[17] Feynman and Hutchings, *Surely you're joking, Mr. Feynman!*, p. 338.
[18] Feynman and Hutchings, *Surely you're joking, Mr. Feynman!*, p. 338.
[19] N. Oreskes, *Why Trust Science?* (Princeton University Press, 2019), p. 247.
[20] Ibid.
[21] N. Oreskes, "Scientists Get Things Wrong. But We Should Still Trust Science," Time.com, 2019, Accessed: May 28, 2021, https://time.com/5709691/why-trust-science/

None of this is to say that we should not "trust science" at least in one sense. For centuries, science has proven itself to be a worthy pursuit, and scientific claims are often trustworthy. The scientific method has earned trust, at least in my view. Where Oreskes goes awry, however, is with the traffic-cop-esque "nothing to see here" hand-waiving. We are told to pay no attention to that man behind the curtain. A "consensus" has been reached don't you see, so move along. While Oreskes is right that you wouldn't want to perform your own open-heart surgery or grow your own wheat to make bread, that is a bit of a mixed metaphor for the truth claims of science. You do not need to know optimal brewing techniques to buy and enjoy good-quality beer. The flour and sugar in the pantry do not make any demands on your mind. The truth-claims of the scientist do, however. The products of science are not consumable goods; they are assertions of truth. A scientist's demand that I believe something as true is nothing like the plumber's demand that I delegate the task of fixing my faucet to him.

While science should be trusted, there is nothing wrong with the People taking a peek behind the curtain before giving any particular, purportedly scientific, *claim* their own personal assent. Ronald Reagan's adviser on Russian affairs, Suzanne Massie, taught him the Russian proverb "doveryai no proveryai"—"trust, but verify." All humans are subject to the Idols of the Tribe, and there are always one or more humans behind the curtain pulling the levers on any particular chunk of "scientific knowledge." Yes, trust science and the scientific method, but the People can maintain their liberty to verify science's human outputs. Indeed, sometimes laypeople are actually in the best position to identify scientific lapses caused by the Idols of the Tribe.

In 1998, for example, thirteen scientists published a paper in the highly-prestigious and peer-reviewed U.K. medical journal, *The Lancet*.[22] Like most papers published in medical journals, it was obtusely titled: *Ileal-lymphoid-nodular hyperplasia, non-specific colitis, and pervasive developmental disorder in children*. The basic findings of the paper were quite clear, however. This baker's dozen of scientists led by Andrew Wakefield found that "[o]nset of behavioral symptoms," such as autism, were "generally associated in time" with the administration of the "measles, mumps, and rubella vaccination."[23] Of course, these findings made international news. The UK's Medical Research Council convened a special meeting to discuss the findings a month after their publication. Wakefield maintained the scientific integrity of his study and told

---

[22] Rao and Andrade, "The MMR Vaccine and Autism: Sensation, Refutation, Retraction, and Fraud," *Indian J. Psychiatry* 53, 2 (2011): pp. 95-96 (summarizing relevant events).
[23] A. Wakefield et al., "Ileal-lymphoid-nodular hyperplasia, non-specific colitis, and pervasive developmental disorder in children," *The Lancet* 351, 9103 (1998): pp. 637-41.

the Council that all of the parents and children for the study had come to his group "without any connection through any other organization."

But despite clearing the peer-review process of a highly prestigious medical journal, the suspicious nature of the study is apparent even on its face. The published paper acknowledged that the entire study was conducted on the basis of only twelve children who received the measles, mumps, and rubella vaccine. One hundred percent of the children had "intestinal abnormalities."[24] Eleven of the children suffered from chronic inflammation of the colon. Nine of them were autistic, and one had "disintegrative psychosis."[25] Based on such a sample, it is apparent on its face that no honest scientist should be able to draw a conclusion about the measles, mumps, and rubella vaccine's impact on the children. There was clearly no control, and the monolithic similarities in the children was obviously contrived.

And yet, thirteen scientists and a prestigious medical journal put their name to the paper. Instead of dismissing the study's claims out of hand, the scientific community convened special meetings to discuss the findings. Wakefield was interviewed in special reports of the BBC and put on 60 Minutes. While some other studies were published with data suggesting no link between the vaccine and autism, it was left to the People to finally discredit the paper. In 2004, the investigative journalist Brian Deer reported in *The Sunday Times* that, contrary to Wakefield's claims to the Medical Research Council, some of the parents of the twelve children in the study were referred to Wakefield's group by a UK lawyer that wanted to file a lawsuit against vaccine manufacturers. Wakefield and the hospital where he worked had also received funding from the lawyer and the UK's Legal Aid Board respectively. Only after this report did *The Lancet*'s editor, Richard Horton, acknowledge that the 1998 paper was "fatally flawed" and should have been rejected as biased. (*The Lancet* itself did not public a formal retraction of the paper until 2010.) And only after Deer's report did ten of Wakefield's co-authors issue a partial retraction directed to an "interpretation" of the paper.

While the salacious details of Wakefield's study simplified the discrediting process, the Idols of the Tribe were on full display the whole time: a tiny and homogenous sample with suspicious data grouping correlating the precise parameters the scientists wanted to study (the vaccine, intestinal disorders, and behavioral disorders). Bacon had explicitly warned against the tendency to

---

[24] Wakefield et al., "Ileal-lymphoid-nodular hyperplasia, non-specific colitis, and pervasive developmental disorder in children," pp. 637-41.

[25] Wakefield et al., "Ileal-lymphoid-nodular hyperplasia, non-specific colitis, and pervasive developmental disorder in children," pp. 637-41.

associate things that strike the senses closely in time. But this is precisely what the thirteen authors of the paper did, interpreting the data based on the children's developmental regression being "generally associated in time with possible environmental triggers."[26] The alternative explanation that the study made no effort to control was staring them in the face: the time that children customarily get the vaccine happens to occur somewhat before symptoms of autism tend to manifest. Add to all this the fact that Wakefield had already formed his anti-vaccination theories. Three years before his 1998 publication, Wakefield had published another paper in *The Lancet* proposing that there was a link between the measles vaccine and Crohn's disease.

The Idols of the Tribe were indulged to the maximum, as data was used to fit a theory. It was a disappointing chapter in such research, one whose impacts are still being felt and suffered to this day. It would be easy to dismiss the Wakefield incident as an outlier of quackery. The scientific community has, for good reason, disowned him. But while the Wakefield case is a particularly egregious and high-profile example on a political hot topic, there is evidence that the Idols of the Tribe are alive and well much more broadly today.

In recent decades, the field of "metascience" has become increasingly prominent. This field involves an effort to turn the scientific method in on itself, to do research on science's research, and to use science to analyze science. Metascience's findings have been shocking. Especially in social sciences like psychology, the growing consensus is that broad swaths of the scientific community are encountering a "replication crisis."

In 2012, Harold Pashler and Eric-Jan Wagenmakers edited a special edition in the journal *Perspectives on Psychological Science* on the "crisis of confidence" confronting practitioners based on the replicability of psychological science. In their introduction to this edition, the editors described the "unprecedented level of doubt among practitioners about the reliability of research findings in the field."[27] They lamented the "hypercompetitive academic climate and incentive scheme that provides rich rewards for overselling one's work and few rewards at all for caution and circumspection" or for engaging in work trying to replicate other published studies. Moreover, even the hard sciences had been impacted. In 2012, for example, pharmaceutical companies revealed that their "efforts to replicate exciting preclinical findings from published academic studies in cancer

---

[26] Wakefield et al., "Ileal-lymphoid-nodular hyperplasia, non-specific colitis, and pervasive developmental disorder in children," pp. 637-41.
[27] Pashler and Wagenmakers, "Editors' Introduction to the Special Section on Replicability in Psychological Science: A Crisis of Confidence?" *Perspectives on Psychological Science* 7, 6 (2012).

biology were only rarely verifying the original results." They warned that, as the articles in the special edition demonstrated, these problems "will not be so easily overcome, as they reflect deep-seated human biases and well-entrenched incentives that shape behavior of individuals and institutions."[28] In other words, the Idols of the Tribe have been allowed to run amuck.

Pashler and Wagenmakers have hardly been alone in ringing the alarm bells. One foundational and well-cited paper in the field of metascience published by physician John Ioannidis is aptly titled: *Why Most Published Research Findings Are False.*[29] Ioannidis used mathematical modelling to demonstrate that in the many fields of modern science that often ignore result replication, it can be proven that "most claimed research findings are false." He accordingly decried the "high rate of nonreplication (lack of confirmation) of research discoveries" and the prevalent strategy of "claiming conclusive research findings solely on the basis of a single study assessed by formal statistical significance, typically for *p*-value less than 0.05." Ioannidis believes that many published findings result from bias manifested in a "manipulation in the analysis or reporting of findings [such as] selective or distorted reporting." His modelling was also able to show that, "paradoxically," "[t]he hotter a scientific field (with more scientific teams involved), the less likely the research findings are to be true." Instead, he saw such fields characterized by a "Proteus phenomenon" where the published results exhibited statistically unlikely patterns in "rapidly alternating extreme research claims and extremely opposite refutations."[30]

In 2018, meta-science researchers Chris Allen and David Mehler published a paper that attempted to quantify the impact of at least one type of Idol of the Tribe.[31] Citing a wide array of other meta-scientists, they concurred with the observation that "[p]ervasive failures to replicate published work have raised major concerns in psychology and other disciplines, which [have] been termed a 'crisis.'" As one potential remedy, they suggested that scientists should be required to publicly declare their hypotheses and the mechanism by which they propose to analyze data before they began experiments. This prophylactic measure would hopefully stymie the human tendency for "post-hoc exploratory analyses" of data to achieve a preferred result. To test this, the authors surveyed a significant sample of published bio-medical and psychological papers that used

---

[28] Pashler and Wagenmakers, "Editors' Introduction to the Special Section on Replicability in Psychological Science: A Crisis of Confidence?"
[29] J. P. A. Ioannidis, "Why Most Published Research Findings Are False," PLoS Med 2, 8 (2005).
[30] Ibid.
[31] C. Allen and D. Mehler, "Open Science Challenges, Benefits and Tips in Early Career and Beyond," PsyArXiv (17 Oct. 2018).

such methodological "pre-registrations," and found that 61% of these studies concluded that the data did not support the hypothesis. In contrast, only 5-20% of papers published in the "traditional" scientific literature conclude that their hypotheses were not supported by the experimental data.[32]

As these examples and the recent findings of metascience show, Bacon's Idols of the Tribe have exerted an enduring and resilient influence on science. Human nature does not change. A scientist does not have to be badly motivated or consciously unethical for scientific results to be unwittingly impacted. And sometimes it takes a fresh perspective from an outsider—a Brian Deer—to see the human biases and mistakes in the facially dispassionate and objective language of published scientific papers. Above all, the People should encourage science performed with integrity and humility, where, in Feynman's language, the scientist bends over backgrounds to candidly disclose all the weaknesses in an experiment and all of the ways a theory can be wrong. True confidence is seen when the scientist unabashedly exposes the vulnerabilities of his or her work, not in the cagey, chip-on-your-shoulder, aloofness that we often see from those like Naomi Oreskes who believe it is a bad thing when science is questioned. If a theory is going to stand the test of time, a scientist should relish second-guessing, respond to criticism with "do your worst," and be glad to be proven wrong for the sake of the Truth.

At the end of his highly successful career, the Nobel Prize-winning physicist Heinrich Rohrer knew better than most how the Idols of the Tribe ply their terrible work on even the very best scientists. In a lecture in Stockholm about a year before his death, he asked the scientific community to give young scientists a fresh start: "we must be careful not to corrupt their work with the questionable practices that the scientific community has adopted in recent years. The new generation of researchers must be given the skills and values — not just scientific ideals, but also awareness of human weaknesses — that will enable it to correct its forebears' mistakes."[33] These human weaknesses must be overcome not only on the level of the "tribe," but, as discussed next, on the individual level as well.

---

[32] Allen and Mehler, "Open Science Challenges, Benefits and Tips in Early Career and Beyond."
[33] H. Rohrer, "The Misconduct of Science?," *Proceedings of Trust, Confidence, and Scientific Research*, 2021, Accessed: May 28, 2021, http://www.abc.net.au/science/articles/2012/07/16/3546732.htm?site=science_dev&

The Idols of the Cave

The idols of the cave "derive their origin from the peculiar nature of each individual's mind and body, and also from education, habit, and accident."[34] They can arise where the scientist becomes too "attached" to a particular field or "contemplation," either from having made a significant contribution to it or having "bestowed the greatest pains upon such subjects, and thus becom[ing] most habituated to them." Such idols can "wrest and corrupt" a scientist through their "preconceived fancies." A scientist may also "evince unbounded admiration" for certain authorities, viewpoints, or outcomes.[35]

However, Bacon advised that "truth [should] not be sought in the good fortune of any particular conjecture of time, which is uncertain, but in the light of nature and experience, which is eternal…In general, he who contemplates nature should suspect whatever particularly takes and fixes his understanding, and should use so much the more caution to preserve it equable and unprejudiced."[36]

Idols of the cave have their greatest impact on the social sciences because they embrace the greatest implications for individual predispositions, personal opinions, and prejudices. Scientists' conclusions drawn from their study of history, economics, psychology, sociology, *etc.* more often have far greater implications for morality, politics, and religious views than the results of physics or chemistry.

Even if primarily at play in the social sciences, there are some conclusions even in physics whose political or moral appeal may differ depending on the particular outlook of the scientist. The Big Bang theory, for example, was originally proposed by a scientist-priest, Father Georges Lamaitre, in 1927. In 1951, the Pope welcomed the idea of the Big Bang in an address to the Pontifical Academy of Sciences, claiming that "present-day science, with one sweeping step back across millions of centuries, has succeeded in bearing witness to that primordial *fiat lux* uttered at the moment when, along with matter, there burst forth from nothing a sea of light and radiation." On the other hand, British physicist William Bonner was suspicious of the theory, claiming that it was only motivated "to bring in God as creator" in order to undo what he thought science was supposed to do in "depos[ing] religion from the minds of rational men."[37] Similarly, the astronomer Fred Hoyle condemned the Big Bang theory as being based on Judeo-Christian religion. He continued to argue that the universe was

---

[34] Bacon, *The Novum Organum*, at LIII.
[35] Bacon, *The Novum Organum*, at LIII.
[36] Bacon, *The Novum Organum*, at LIII.
[37] S. Singh, *Big Bang* (Milan: Mondolibri, 2005), p. 361.

eternal and infinite, claiming that the expansion of the universe was offset by a hypothesized continual creation of matter and energy.[38] Further experiments in the mid-Twentieth Century proved these counter hypotheses to the Big Bang wrong.

Moreover, it is not just a scientist's preference for hypotheses that can be impacted by the idols of the cave. Their results can also be impacted, even in the physical sciences. Robert Andrews Millikan was one of the towering figures of early Twentieth Century experimental physics. In 1923, he received the Nobel Prize in Physics based on his measurement of the charge of an electron and Plank's constant. Millikan's design for the experiment to measure the charge of an individual electron was elegant and "has been called one of the most beautiful in physics history."[39] The experiment involved passing tiny oil droplets through charged metal plates. Millikan compared the rise or fall rates of charged droplets against those of uncharged droplets, and discovered that the implied charge values of the droplets were all multiples of a single elementary charge—the charge of a single electron.

> Millikan began these experiments in 1909 and reported his first batch of results in 1910. A Viennese physicist, Felix Ehrenhaft, claimed to have conducted a similar experiment, measuring a much smaller value for the elementary charge. Ehrenhaft argued that his data supported his theory of the existence of "subelectrons." This challenge from Ehrenhaft prompted Millikan to improve his experiment and collect more data to prove that he was right. He improved his experiment and collected more data, which he published in 1913, claiming that his measurement had only a 0.2% uncertainty and disproving the existence of Ehrenhaft's subelectrons. Millikan's measurement that an electron carried a charge of 1.592×10^{-19} coulombs lead, in large part, to his Nobel Prize ten years later.

After Millikan received the Nobel, other scientists looked to recreate Millikan's experiment and measurements. Today, it is accepted that an electron's charge is 1.602×10^{-19} coulombs, about a 0.6% difference from Millikan's measurement (and outside of Millikan's claimed error rate). Based on the real value of this physical constant, one would expect that the data collected by subsequent scientists would reflect Millikan's error. But what occurred was quite different. The later published values only slowly increased towards the correct value, rather than reflecting the random distribution of results centered around the

---

[38] R. Sheldrake, *Science set free* (New York: Deepak Chopra Books, 2013), p. 65-66.
[39] E. Tretkoff, "This Month in Physics History: August 1913: Robert Millikan Reports His Oil Drop Results," *American Physical Society* 15, 8 (2006).

1.602 figure that you would statistically expect. For example, in 1929, Erik Bäcklin published his results in the highly respected publication *Nature*. He reported a measurement of 1.599×10^{-19} coulombs, with uncertainty limits that encompassed Millikan's value.[40] Richard Feynman discussed this history in his *Cargo Cult Science* address, saying:

[Millikan's measurement is] a little bit off because he had the incorrect value for the viscosity of air. It's interesting to look at the history of measurements of the charge of an electron, after Millikan. If you plot them as a function of time, you find that one is a little bit bigger than Millikan's, and the next one's a little bit bigger than that, and the next one's a little bit bigger than that, until finally they settle down to a number which is higher. Why didn't they discover the new number was higher right away? It's a thing that scientists are ashamed of—this history—because it's apparent that people did things like this: When they got a number that was too high above Millikan's, they thought something must be wrong—and they would look for and find a reason why something might be wrong. When they got a number close to Millikan's value they didn't look so hard. And so they eliminated the numbers that were too far off, and did other things like that.[41]

In fact, Feynman himself may have been pulling his punches on Millikan. There are many who believe that Millikan may have allowed Ehrenhaft's challenge to his conclusions get the better of him, causing an otherwise excellent scientist to pay homage to some of the Idols of the Tribe. Millikan's 1913 paper claimed, with respect to his reported data on 58 oil drops: "this is not a selected group of drops, but represents all the drops experimented upon during 60 consecutive days."[42] Later scientists, however, obtained Millikan's laboratory notebooks and found that approximately 75 drops had been measured in that approximate timeframe, and Millikan had editorialized about his results in the margins of his unpublished notebooks.[43] He wrote thing like "publish this beautiful one," "error high will not use," "perfect publish," "won't work," "too high by 1 ½ %," and "too high e by 1 1/4 %."[44]

---

[40] E. Bäcklin, "Eddington's Hypothesis and the Electronic Charge," *Nature* 123 (1929), pp. 409–410.
[41] Feynman and Hutchings, *Surely you're joking, Mr. Feynman!*, p. 338.
[42] Tretkoff, "This Month in Physics History: August 1913: Robert Millikan Reports His Oil Drop Results."
[43] Tretkoff, "This Month in Physics History: August 1913: Robert Millikan Reports His Oil Drop Results"; *see also* http://yclept.ucdavis.edu/course/280/Millikan.pdf
[44] Tretkoff, "This Month in Physics History: August 1913: Robert Millikan Reports His Oil Drop Results."

This is an interesting example where the Idols of the Tribe and the Idols of the Cave operated in tandem, in a case of pure physical science where it is hard to imagine a scientist having any philosophical or political preference for a number, *i.e.*, who really cares if the number is 1.592 versus 1.602? And the point here is not to vilify and judge Millikan or the scientists who did the later measurements. The point is that they are humans. At least as it relates to the Idols of the Cave, it was the very fact that the later scientists saw Millikan as super-human that led them astray. If scientists are susceptible to awe of scientific authority, how much more on guard should the People be?

To be sure, the Millikan example is not an isolated incident, even in physics. The speed of light, c, has been a fundamental constant of physics at least since Einstein's theory of special relativity. "[E]arly measurements of the speed of light varied considerably, but by 1927, the measured values had converged to 299,796 kilometers per second."[45] In 1929, leading American physicist, Raymond Birge, published a treatise on the physical constants, claiming that "[t]he present value of c is entirely satisfactory and can be considered more or less permanently established."[46] Nonetheless, from 1928 to 1945, the measurements of the speed of light dropped by 20 kilometers per second, and the values measured by scientists were in impressively close agreement with each other. Then, in the late 1940s, the accepted speed of light went up again by 20 kilometers per second, returning to the 1920s values, and a new consensus developed around that figure. In 1985, the British metrologist (metrology is the scientific study of measurement), explained what to this point should be clear:

> The tendency for experiments in a given epoch to agree with one another has been described by the delicate phrase "intellectual phase locking." Most metrologists are very conscious of the possible existence of such effects; indeed ever-helpful colleagues delight in pointing them out!...[T]he accusation that one is most likely to stop worrying about correction when the value is closest to other results is easy to make and difficult to refute.[47]

If even the physical sciences are beset by the Idols of the Cave, their influence on the social sciences has been many times greater. Innumerable such examples can be given. But, to get a broad view, one example might be drawn

---

[45] Sheldrake, *Science set free*, p. 92.
[46] R. Birge, "Probable Values of the General Physical Constants," *Review of Modern Physics* 33 (1929): pp. 233-39.
[47] B. Petley, *The Fundamental Physical Constants and the Frontiers of Metrology* (Adam Hilger, 1985), p. 294.

from a field where a scholar's intellectual, political, and religious pre-commitments are near their pinnacle—the historical study of the Bible.

Historical criticism of Biblical texts has some roots in Enlightenment skepticism. The philosopher Baruch Spinoza, for example, popularized the argument that Moses could not have written the Torah because a part of the Torah describes Moses' death and events occurring immediately thereafter. But the fledgling science of biblical historical criticism really began in earnest in the Nineteenth Century, during the Romantic Period. The "roots of historical criticism" were "meant to be value-neutral, or disinterested."[48] This new science "tried, so far as possible, to approach the text without prejudice, and to ask not what it meant 'for me,' but simply what it meant…The historical critic's calling was to be a neutral observer, prescinding from any kind of faith commitment in order to get at the truth."[49] This initial period of scholarship reached its culmination (at least in terms of its impact on popular culture) with Albert Schweitzer's 1906 bestselling book *The Quest for the Historical Jesus*, where he concluded that the traditional view of Jesus had been "designed by rationalism" and "endowed with life by liberalism" to create a Jesus that "never existed" and whose image had now "fallen to pieces" in light of science.[50]

While Twentieth Century experts now reject many of Schweitzer's arguments and conclusions, the trajectory of the scholarship remained the same. After another hundred-plus year of non-ending scholarship (including second and third "quests" for the historical Jesus, the Jesus Seminar, *etc.*), the proposed impact of the scholarship remains the same. Scholars often assert that people "need to realize that historical criticism demands (and should be given) authority to qualify and adjust their historical and ultimately theological convictions, beginning with their understanding of the Bible."[51] As noted Biblical scholar Bart Ehrman observed, the scholars have "investigate[d] the Gospels of the New Testament with a critical eye to determine which stories, and which parts of stories, are historically accurate with respect to the historical Jesus, and which represent later embellishments by his devoted followers."[52] As a result, according to these scholars, lay people should

---

[48] S. Hahn and B. Wiker, *Politicizing the Bible: The Roots of Historical Criticism and the Secularization of Scripture*, (Germany: Herder & Herder, 2013), p. 1.
[49] Ibid.
[50] A. Schweitzer, W. Montgomery, and C. Black, *The Quest for the Historical Jesus* (1910).
[51] R. Yarbrough, "Should Evangelicals Embrace Historical Criticism? The Hays-Ansberry Proposal," *Themelios* 39, 1 (2014): pp. 37-52.
[52] B. Ehrman, *How Jesus Became God* (New York: HarperOne, 2015), p. 13.

acknowledge this "compelling and intriguing field of study."[53] And yet, these scholars find it "striking" that most readers "know almost nothing about textual criticism" and accordingly do not even realize that "there is even a 'problem' with the text" or that "through the application of some rather rigorous methods of analysis, [scholars have] reconstruct[ed]" the original text.[54] In other words, they say: why can't religious people just "follow the science" and see that many tenets of their faith have been supposedly proven false by science?

After the Supreme Court decided the *Hobby Lobby* case about religious freedom, an article published in Salon summarized the political upshot of the supposed Biblical in-knowledge of science. A "staggering majority of Americans...have not a morsel of knowledge as it pertains to just about all aspects of historical context and biblical scholarship....Knowing the Bible requires a scholarly contextual understanding of authorship, history and interpretation."[55] This is all just to say that the science of historical biblical criticism has real-world political stakes. But what Idols of the Cave might be lurking beneath the surface of this centuries-old line of scholarship?

A good example for our purposes can be drawn from the "science" of dating when the various books of the Bible were written. While such questions might, at first blush, seem to be relatively pedantic, they have many important theological implications. Claims that much of the Torah was not written during the general epoch that Moses lived are used to argue that the ancient Israelites may not have been monotheists. And with respect to the New Testament, scholars use a late dating of certain books to argue that they have less historical accuracy and, instead, reflect later inventions of the events in order to deal with the issues faced by the early church. In that regard, it is generally believed that the fourth gospel, the Gospel of John, was the last of the four gospels of the New Testament to be written. But can the science of historical criticism tell us when exactly it was written (at least as exactly as possible) and evaluate the implications of such dating?

Most of the early Nineteenth Century historical-critical scholarship on this issue was done in Germany. The Hegelian German philosopher Ferdinand Christian Bauer was an influential leader in this early work and founded the Tübingen School of Theology. He argued that the Gospel of John had to be

---

[53] B. Ehrman, *Misquoting Jesus: The Story Behind Who Changed the Bible and Why* (New York: HarperOne, 2007), p. 14-15.
[54] Ibid.
[55] C. J. Werleman, "How America's Biblical ignorance enables the Christian right," Salon, 2014, Accessed: May 28, 2021, https://www.salon.com/2014/07/10/how_americas_ignorance_about_the_bible_allowed_the_christian_right_to_push_its_extreme_agenda_partner/

heavily influenced by later Greek and gnostic philosophies and, accordingly, believed that it had been written around AD 170.[56] In 1820, a contemporary of Bauer, Karl Gottlieb Bretschneider published a highly influential treatise on the Gospel of John collecting the alleged evidence for a late dating of John's Gospel. The "dominating interest" of the work is "to prove that the date and the origin of the gospel are incompatible with the apostle John. The real author, [according to] Bretschneider, was an anonymous Christian philosopher who wrote about the second decade of the second century. Earlier than that, no trace of the gospel can be found."[57]

The late dating of the Gospel of John remained the expert consensus for many decades and became so widely accepted and embedded in the scholarship so as to have become mainstream. In 1925, a Catholic priest named Joseph Turmel (writing under the pseudonym Henri Delafosse) published a book arguing that the Gospel of John originated in a heresy of the early Church called Marcionism and dated it to AD 170-175.[58] In 1936, former priest and scholar Alfred Loisy estimated that the Gospel of John had been written in AD 150-160, and based on the overwhelming weight of the scholarship claimed that "the first publication can hardly have been effected before 135-40."[59]

It is likely that this supposedly "scientific" consensus would have remained the same through today were it not for cataclysmic shift caused by the chance discovery of a scrap of ancient papyrus in Egypt. This fragment, measuring only about 3.5 inches by 2.5 inches is known as Rylands Library Papyrus P52 and may be the earliest surviving copy of the New Testament that modern archeology has been able to discover. The contents of P52 were first published in 1935 by Colin Roberts, who dated it to the "first half of the second century," finding that it very closely resembled other ancient papyri from the late first and early second centuries, including a scroll containing part of Homer's Iliad from that time.[60]

P52 is an excerpt from the Gospel of John. For a variety of reasons now accepted by the majority of scholars, this scrap of papyrus meant that the Gospel of John had to have been first published in the first century, likely around AD 90. As one scholar put it, despite the Nineteenth Century consensus

---

[56] D. Wallace, "John 5,2 and the Date of the Fourth Gospel," *Biblica* 71, 2 (1990): p. 177.
[57] J. Moffatt, "Ninety Years After: A Survey of Bretschneider's "Probabilia" in Light of Subsequent Johannine Criticism," *The American Journal of Theology* 17, 3 (1913): p. 375.
[58] H. Delafosse, *Le Quatrieme Evangile* (Paris, 1925), p. 127.
[59] A. Loisy, *The Origins of the New Testament* (France, 1936), p. 193.
[60] C. H. Roberts, *An Unpublished Fragment of the Fourth Gospel in the John Rylands Library* (Manchester University Press, 1935).

The False Idols   105

of a much later date, Collin Robert's publication of P52 "changed the boundaries of dating the FG [i.e., Fourth Gospel] forever."[61] Indeed, the discovery of P52 was "soon followed by the discovery of Papyrus Egerton 2," which has also "generally been dated in the first half of the second century." Papyrus Egerton 2 comments on the gospels and quotes from the Gospel of John, which "provides an indication that by the time of its writing the [Gospel of John] was regarded as equally authoritative as the [other gospels]" and "clearly indicates that John had not just been written."[62]

While these chance discoveries precluded the experts' claims that John's Gospel was written in the context of later second century heresies, the Idols of the Cave are tenacious and still persevered in their influence. Scholars tried to keep much of political impact of the Nineteenth Century theories while acknowledging that John's gospel was written much earlier. In the 1960s and 70s, for example, scholars hypothesized the existence of a "Johannine community" out of which the Gospel of John arose. This hypothetical community had separated itself from the other early Christian communities. As one scholar put it, the Johannine community was a "distinct trajectory" in early Christianity that had "estranged" itself "from the wider society, the society of the synagogue, and even the society of other Christian groups."[63] But again, as time wore on, this emperor too was found to lack clothes. In the most recent decades, a new generation of experts saw that this hypothesis was "entirely a scholarly construct, the product of a circular hermeneutical process" and "historical fiction."[64] The new consensus saw that the Gospel of John was not as different from the other early Christian writings and not "estranged" from the Jewish origins of Christianity. "If we attend to the way scripture actually functions in John, we will see that the identity of Jesus is deeply embedded in Israel's texts and traditions...[I]t is impossible to understand John's Jesus apart from the story of Israel."[65]

The purpose here is not to engage exhaustively with this extensive line of scholarship. And it is certainly not to lay out a case for what I think were circumstances under which this ancient document was written. It is to give a glimpse of the tenacious workings of the Idols of the Cave in an area where a scholar's viewpoints, politics, religion, and preconceived notions are heightened. In the Nineteenth Century, the science was *sure* that this document was written

---

[61] D. Croteau, "An Analysis of the Arguments for the Dating of the Fourth Gospel," *Liberty University Faculty Publications* Paper 118 (2003): p. 60.
[62] Ibid.
[63] D. Lamb, *Text, Context and the Johannine Community: A Sociolinguistic Analysis of the Johannine Writings* (A&C Black, 2014), p. 1 (quoting Stephen C. Barton).
[64] Ibid. (quoting Adele Reinhartz and Edward Klink).
[65] R. Hays, *Echoes of Scripture in the Gospels* (Baylor University Press, 2016), p. 287.

in the mid-second century in the context of certain political controversies of the early Christians...until it wasn't. P52 proved that consensus was absolutely wrong. But then the science became *sure* that it was written by an estranged sect in the late first century who had broken away from Judaism and the other Christians...again, until it wasn't.

This is not to say that historical scholarship should be ignored or ended—far from it. It is just a call to recognize how intensely the Idols of the Cave can work upon the very human individuals that publish such scholarship. As one religious leader put it, the various "reconstructions of [the historical] Jesus...are much more like photographs of their authors and the ideals they hold."[66] Or as a duo of scholars concluded after conducting a meta-scientific review of the history of such scholarship (in much greater detail than anything that could be done here): "[We have] every reason to believe that significantly more detailed studies of the politicizing aspects of nineteenth-century scriptural scholarship are called for, and only such studies can hope to disentangle the legitimate tools of the historical-critical method from the various political and secular aims."[67]

The Idols of the Cave are alive and well in science today.

The Idols of the Marketplace

Idols of the Marketplace are those threats to science that arise from "the intercourse and association of men [and women] with each other."[68] For Sir Francis Bacon, the primary idol of the marketplace was the misuse of language in science. During his time, science was still emerging from the medieval era, and it drew its language from religious backgrounds and the early efforts of quasi-scientific endeavors like alchemy. Accordingly, Bacon lamented the loose use of language like ether, the "element of fire," and fortune. Single words like "humid" were being used to describe a wide range of phenomenon just because of a perceived similarity in quality. For example, materials with low viscosity and materials with a low melting point were both described as humid.

The periodic table of elements had not been discovered when Bacon wrote. Nor was most science done within a mathematical framework. Sir Isaac Newton revolutionized science toward the end of the Seventeenth Century, half a century after Bacon's death, with his *Philosophiae Naturalis Principia Mathematica* (Mathematical Principles of Natural Philosophy). And from that point on, every

---

[66] J. Ratzinger, *Jesus of Nazareth: From the Baptism in the Jordan to the Transfiguration* (Doubleday, 2007), p. xi-xii.
[67] Hahn and Wiker, *Politicizing the Bible: The Roots of Historical Criticism and the Secularization of Scripture*, p. 565-56.
[68] Bacon, *The Novum Organum*, at XLIII.

major leap forward in the physical sciences was based primarily in a mathematical system: the field theories of Maxwell's electromagnetism, Einstein's special and general relativity equations, Schrodinger's wavefunction equation for quantum mechanics.

The social sciences envied the confidence and certainty afforded by the mathematics of the physical sciences. And the social sciences developed, they became more and more characterized by their own forms of mathematical expression. The catchphrases for communicating data in the social sciences became "p values" and standard deviations. "Statistical significance" became the watch word. It was here that the Idols of the Marketplace re-emerged as scientific results became increasing reported mathematically through statistical analysis.

In his autobiography, Mark Twain reported that he was perplexed that, as he become older, he seemed to write fewer and fewer words per day and yet he felt that his literary output had remained the same. Recalling the old adage that "[t]here are three kinds of lies: lies, damned lies, and statistics," Twain concluded that "[f]igures often beguile me, particularly when I have the arranging of them myself."[69] As is often the case with clichés, this one laid its finger on a true insight.

Jacob Cohen was an American psychologist and statistician that became concerned with the way statistics were being used, particularly in the social sciences, to make unsupported truth claims. One of his foundational papers, *The Earth is Round (p<.05)*, lamented how, despite decades of criticism from statisticians, "the ritual of null hypothesis significance testing—mechanical dichotomous decisions around a sacred .05 criterion—still persists."[70] Recalling *The Emperor's New Clothes*, Cohen believed that "this naked emperor" in statistical analysis "has been shamelessly running around for a long time."[71]

For decades before Cohen's paper and for decades since, scientific papers have analyzed data more or less as follows. The scientist comes up with a hypothesis. For illustrative purposes, let's say the hypothesis is that attending a full-time Pre-K program will lead to a higher salary later in life, on average, than attending a part-time Pre-K. If that is the hypothesis, then the "null hypothesis" is that attending a full-time Pre-K program will not lead to a higher salary later in life, on average, than attending a part-time Pre-K. The scientist then goes and collects an adequately large sample of data, for example, by surveying adults to learn their salaries and whether they attended a full-time Pre-K, a part-time

---

[69] M. Twain, "Chapters from My Autobiography," *North American Review*, 1906, Accessed May 28, 2021, http://www.gutenberg.org/files/19987/19987.txt
[70] J. Cohen, "The Earth is Round (p<.05)," *American Psychologist* 49, 12 (1994): pp. 997-1003.
[71] Cohen, "The Earth is Round (p<.05)," pp. 997-1003.

Pre-K, or no Pre-K. Hopefully, adequate controls will be put into place regarding variables in the data, such as the respondents' age and geographic location. But once those controls are added, the scientist will perform a statistical analysis on the data to test the null hypothesis. The "sacred .05 criterion" mentioned by Cohen is the generally accepted criterion that the data has established something "statistically significant" if the likelihood of the null hypothesis in view of the data is 5% or less, i.e., $p<.05$. Of course, this all appears perfectly logical and scientific.

However, Cohen reminded the scientific community of the "permanent illusion" in the way people misinterpret such statistical testing. Other statisticians called it the "Bayesian Id's wishful thinking." The problems with the .05 criterion are "its near-universal misinterpretation of p as the probability that [the null hypothesis] is false, the misinterpretation that its complement is the probability of successful replication, and the mistaken assumption that if one rejects [the null hypothesis] one thereby affirms the theory that led to the test."[72] In other words, if the scientist in the Pre-K experiment came up with a p-value less than .05, then many scientists and almost all lay news-reporters are going to report the study as showing that attending a full-time Pre-K leads to higher salaries later in life. But that is simply not what the statistics have shown.

Cohen analogized the logic that leads to such a sweeping conclusion based on a $p<.05$ criterion to the following argument:

1. If a person is an American, then he is probably not a member of Congress.
2. This person is a member of Congress.
3. Therefore, he is probably not an American.

Premise (1) is obviously true, and true to a much greater extent than $p<.05$. Premise (2) is the data. However, this example illustrates how the logic of the argument does not hold. Cohen points out that the logic of the misinterpretation of scientific data relies on the exact same form of argument, namely:

1. If the null hypothesis is true, then this result [the data] would probably not occur.
2. This result [the data] has occurred.
3. Therefore, the null hypothesis is probably not true and, therefore, formally invalid.

---

[72] Cohen, "The Earth is Round ($p<.05$)," pp. 997-1003.

And yet, this "formulation appears at least implicitly in article after article in psychological journals" and in other scientific fields as well.[73]

The statistical fallacy is obvious. Null hypothesis testing would be really good at showing that the scientist's hypothesis has not been proven. (But that is not really what the scientist is hoping for.) When a scientist tests the null hypothesis, the scientist is finding the probability that the data could have arisen if the null hypothesis were true. If that probability is less than 5%, then the scientist is justified in concluding that *if the null hypothesis is true,* the data is unlikely. But that is not nearly as interesting as the inverse question. What the scientist and the public is really interested in is the probability that the null hypothesis is true, *given the data.* These are not the same thing!

To cope with this, most statisticians will just assume a correction factor or a normal distribution of data. But that is often not an appropriate assumption.

For example, consider some of the statistical un-intuitiveness of test results for rare diseases. Let's say that you are given a test for the now rare disease of leprosy. You don't think you have leprosy because you don't have any of the symptoms. But to your shock and horror, the lab comes back and says your blood work has come back positive for leprosy. You protest to your doctor, but she says that the test is 99.9% accurate in returning true positives. You go home thinking, "well, I guess I am a leper."

But wait. Your spouse is a statistician, and she points out that only 0.0001% of people in the world today actually have leprosy. And she tells you that you need to consider that fact relative to the false-positivity rate of the test. If only 1 out of 100,000 people have leprosy, but the test returns a false positive in 1 out of 1,000 cases, then it is much more likely that your test result was a false positive than a true positive. It was only the illusion of certainty present in the false positivity rate that failed to take into account the underlying (extremely high) probability of the null hypothesis: that you did not have leprosy.

This statistical illusion has been persistent in imprecise usage of mathematical language, and it is the same illusion present in a lot of scientific literature today. A new Idol of the Marketplace that Bacon never anticipated has been born and continues to plague science, even if no one worries much today about the imprecision in the word "humid" or "base metal."

The surface level certainty of the statistical language found in scientific literature can often be confounded by unstated, unrecognized, and/or uncontrolled variables. One famous example has been studied and debunked in recent years by metascience. A variety of studies had for a long time reported a

---

[73] Cohen, "The Earth is Round (p<.05)," pp. 997-1003.

link between coffee consumption and lung cancer. This finding always flew in the face of the common-sense implication of the fact that coffee is consumed into the stomach and digestive system, not the lungs. (Statistical idols of the marketplace tend to reveal themselves with common sense.) An unstated assumption of these statistical studies was that coffee drinkers and non-coffee drinkers are just as likely to smoke or engage in other activities that are known to increase the risk of lung cancer. The reality, however, is that coffee drinking and smoking are correlated. Nonetheless, as late as 2016, scientists persisted in drawing a coffee-lung cancer link even after this confounding variable has been pointed out. A 2018 meta-scientific analysis, however, showed that the evidence of such a link simply was not there, though it admirably "encouraged" further analyses with still larger samples "to confirm these results."[74]

The problem of confounding variables highlights another persistent problem in that academia and scientific journals tend to prize research that leads to truth claims, rather than research that helps improve research. Richard Fenyman's *Cargo Cult Science* addressed how valuable the latter type of research could be by giving an interesting example. For much of the Twentieth Century, rat-maze experiments were all the rage, but their results often proved difficult to replicate. Feynman observed that one publication in the rat-maze literature from 1937 had been tragically overlooked. The scientist in that experiment had set out trying to train rats to go to the third door in a corridor, but his rats seemed to be able to tell which door was the one that had the food the time before. Feynman discussed the painstaking labors the scientist went through in order to properly control the experiment:

> The question was, how did the rats know, because the corridor was so beautifully built and so uniform, that this was the same door as before? Obviously there was something about the door that was different from the other doors. So he painted the doors very carefully, arranging the textures on the faces of the doors exactly the same. Still the rats could tell. Then he thought maybe the rats were smelling the food, so he used chemicals to change the smell after each run. Still the rats could tell. Then he realized the rats might be able to tell by seeing the lights and the arrangement in the laboratory like any commonsense person. So he covered the corridor, and, still the rats could tell. He finally found that they could tell by the way the floor sounded when they ran over it. And he could only fix that by putting his corridor in sand. So he covered one

---

[74] V. Galarraga and P. Boffetta, "Coffee Drinking and Risk of Lung Cancer—A Meta-Analysis," *American Association for Cancer Research* 25, 6 (2016): pp. 951-957.

The False Idols

after another of all possible clues and finally was able to fool the rats so that they had to learn to go in the third door. If he relaxed any of his conditions, the rats could tell.[75]

After doing all this work, the scientist published a paper describing all of the measures that are needed in order to fool rats in rat-maze experiments. This was important work because the scientist had discovered everything that you needed to do to properly run such tests on rats.

The problem was that no one cared about this study because it did not contain any affirmative findings about rat behavior or psychology. Later experimenters continued to run their rats in the same old way. This is an example of an Idol of the Marketplace. The market for scientific research does not prize this type of research, or research that disproves certain hypotheses, as much as research that purports prove something true about nature or people. As Bacon put it in the Seventeenth Century, "it is the peculiar and perpetual error of the human intellect to be more moved and excited by affirmatives than by negatives; whereas it ought properly to hold itself indifferently disposed towards both alike." This idol has become known today as "publication bias" or the "file drawer effect." Experiments with statistically significant results in favor of a hypothesis or theory are much more likely to be published and more helpful in advancing the career of new scientists. This idol also often bleeds over with the idols of the tribe and cave because the need to publish in an academic environment often leads to biases in favor of achieving the valuable results.

Idols of the marketplace arise out of the way science is communicated and valued. Even if centuries of science have taught us how to avoid the vagaries of language, that error has been displaced by the persistent illusions of statistical language. A role remains for the People to apply common sense and to offer healthy resistance and skepticism instead of being bowled over by the mischiefs of standard deviations and statistical significance.

The Idols of the Theater

Finally, Bacon explained that the Idols of the Theater exist where scientific theories are allowed to become a "stage-plays, representing worlds of their creation" rather than the natural world.[76] In these plays, "you may observe the same thing which is found in the theater of the poets, that stories invented for the stage are more compact and elegant, and more as one would wish them to

---

[75] Feynman and Hutchings, *Surely you're joking, Mr. Feynman!*, p. 338.
[76] Bacon, *The Novum Organum*, at XLIV.

be, than true stories out of history." These scientists might "bestow[] much diligent and careful labour on a few experiments" but thereby are "made bold to educe and construct [large] systems, wresting all other facts in a strange fashion to conformity therewith."[77]

Then, once these systems are put in place with such "intemperance," the Idols of the Theater will typically "leav[e] no way open to reach and dislodge them," rendering the "sciences dogmatic and magisterial." With the creation of artificially boxed-in systems, Bacon lamented that two camps will typically emerge, both of which are misguided. There will be the orthodoxy, which begrudges and retaliates against any attempt to criticize the theatrical narrative. And there will be a skeptical rebellion that resorts to "hostile confutations" and unfair skepticism.[78]

The Idols of the Theater can exert their grip on even the best scientists operating in hard science. Albert Einstein, for example, did not allow the theater of Newtonian physics to control his scientific inquiries. By the early Twentieth Century, cracks in Newton's gravitational equations had formed. For example, they were unable to explain some of the behavior of fast-moving planets, like Mercury which orbits the sun once every 88 days. Einstein's theories of relativity, however, not only explained why Newton's equations worked very well for the data of common experience, but also proved accurate and predictive for fast-moving bodies like Mercury and other phenomena on the extremes of speed and gravitation.

Einstein had also made some of the early discoveries in quantum mechanics. But before long, quantum theory evolved to a place where, philosophically, Einstein found that he was unwilling to go. Certain physicists working the field of quantum mechanics, most notably Niels Bohr and Werner Heisenberg, became known for endorsing the so-called Copenhagen interpretation of the equations of quantum mechanics. Under this interpretation, the implications of these equations were not just a consequence of the mathematics that people chose as their model to explain certain quantum phenomena. Instead, the implications were real and said something strange about physical reality itself. For example, a somewhat simplified implication of the Heisenberg uncertainty principle and Schrodinger's wave-function equation is that subjective knowledge of certain information about a physical system will change its reality.

To scientists committed to an objective, naturalistic view of the physical world, like Einstein, this claim was anathema. Surely that implication of the mathematics of quantum mechanics just showed that scientists' understanding

---

[77] Bacon, *The Novum Organum*, at XLIV.
[78] Bacon, *The Novum Organum*, at XLIV.

of the quantum-mechanical situation was incomplete. In principle, my knowing that something is true or false does not change the physical world, thought Einstein. And so, together with two similarly minded scientists (Podolsky and Rosen), Einstein published a watershed 1935 paper attacking the Copenhagen interpretation, called *Can Quantum-Mechanical Description of Physical Reality Be Considered Complete?*[79] Einstein's idol of the theater is clearly seen in the way the paper advances their argument.

Einstein argued that in order for a physical theory to be complete "every element of the physical reality must have a counterpart in the physical theory." Consistent with his philosophical commitment, Einstein defined the elements of physical reality objectively to mean anything whose physical quantity we can predict with certainty "without in any way disturbing a system."[80] Then, using the equations of quantum mechanics, Einstein showed that it should be possible to prepare a pair of particles that exist in an entangled state. Then, theoretically, even if those particles were separated by light years, a measurement on the first particle would necessarily and instantaneously effect the second particle. But this seems impossible, both as a matter of common sense notions of cause-and-effect as well as formally under the theory of relativity that nothing can be transmitted faster than the speed of light. This problem became known as the EPR Paradox (EPR standing for the last names of the three authors of the paper).

Based on the paradox, Einstein proposed that both particles must have positions and momenta that are, in principle, defined and knowable at all times. It was just that quantum mechanics itself must be incomplete in being unable to provide sufficient predictions. Einstein acknowledged that "[o]ne could object" to the paper's "conclusion" on the ground that his view of reality is not correct. One could only resolve the paradox if, in reality, the particles actually did not have definite and quantified positions and momenta apart from their measurement. Since either position or momentum can be predicted, but not both simultaneously, "they are not simultaneously real." "This makes the reality of [position] and [momentum] depend on the process of measurement." Einstein, however, rejected this view without further explanation as follows: "No reasonable definition of reality could be expected to permit this." Instead, his view was that quantum mechanics was incomplete and that it must be "possible" in principle to

---

[79] A. Einstein, B. Podolsky, and N. Rosen, "Can Quantum-Mechanical Description of Physical Reality Be Considered Complete?," *Physical Review* 47, 777 (1935).
[80] Einstein, Podolsky, and Rosen, "Can Quantum-Mechanical Description of Physical Reality Be Considered Complete?"

discover the theory that would allow reality to be objectively predicted, contrary to the "incomplete" equations of quantum mechanics.[81]

For some time, this dispute between Einstein's view and the Copenhagen interpretation persisted without resolution by empirical evidence. Einstein continued to reject the physical implications of quantum mechanics, especially under the EPR Paradox, as "spooky action at a distance." Over time, however, the evidence mounted that Einstein was wrong. In 2008, Salart et al. published *Testing Spooky Action at a Distance* in Nature, where they showed empirically that quantum entanglement really exists and that a measurement on one particle affects the other instantaneously (or at least faster than the speed of light).[82] Contrary to relativity, these results suggested the existence of "some hypothetical universally privileged reference frame." These results have been replicated again and again. Scientists committed to Einstein's theater or story and against the more mystical implications of a universally privileged reference frame, lodged a smattering of objections against these studies. But additional experiments, controlling for these objections, have further proven the Copenhagen interpretation to be the truth.[83]

Of course, none of this should be taken to disparage Einstein, who in many other ways transcended scientific paradigms and demonstrated a commitment to free-thinking and independence from prior theaters or stories. Indeed, it may be that quantum mechanics may be itself improved, and adherents to the Copenhagen interpretation should be wary of quantum-mechanical theatrics. The point is just that even the very best scientists in history, working in the most objective scientific fields, can be drawn into their own artificial theaters rather than allowing nature to lead them where it may, however weird and unsettling that might be. Einstein so valued his preferred narrative of an objective physical world that it led him to reject a subjective theory just because it was "spooky" within the story of the world he told himself.

If an Idol of the Theater can beguile one of the best scientists of all time operating in the most objective of all sciences, how much more so should we be wary of scientific-theater when it comes to social sciences or sciences that have unfortunately become politicized. The Idols of the Theater have the potential to create even more mischief when they have the opportunity to become, well, theatrical. In certain areas of science, there has developed a certain level of sensationalism and tendency for "headline-grabbing," along

---

[81] Einstein, Podolsky, and Rosen, "Can Quantum-Mechanical Description of Physical Reality Be Considered Complete?"
[82] D. Salart, "Testing Spooky Action at a Distance," *Nature* 454, 861 (2008).
[83] J. Yin et al., "Bounding the Speed of Spooky Action at a Distance," *Phys. Rev. Lett.* 110, 260407 (2013) arXiv: 1303.0614.

The False Idols

with heated, visceral, and almost emotional reactions to dissenting views. (In contrast, no one justifiably mocked Einstein for his life-long commitment against the Copenhagen interpretation even after the contrary scientific consensus formed at the Fifth Solvay Conference on Physics, held in Brussels in 1927.)

Thus, the following will give some examples of the Idols of the Theater at work in environmental or climate science. But the very politicization of these illustrative observations requires disclaimers upfront. Nothing herein is meant to suggest that climate change is not real, nor is it meant as a jibe at any scientists operating in that field, nor does it at all signal agreement with so-called "climate deniers." Far from it. Instead, the following is simply meant to highlight an Idol of the Theater potentially at play in this field in a few forms and instantiations. As Bacon predicted hundreds of years ago, the Idols of the Theater tend to lead to the creation of an entrenched, emotionally committed orthodoxy, on the one hand, and a defiant, overly-skeptical rebellion, on the other. Bacon did not think *either side* was ultimately positive in the service of science. Science should not be theatric in the sense that it should not use mechanisms other than the reporting of empirical truth to either squelch the dissenting voice or uncharitably criticize the orthodoxy. Science is not a cudgel with which to beat the People into submission, nor is scientific uncertainty a get out of jail free card to unthinkingly reject scientific findings.

And so, with those disclaimers stated and standing, what potential Idols of the Theater might exist in modern science outside of physics?

Some can be seen with retrospection looking at some of the more outlandish predictions and breathless headline-grabs issued by scientists over the decades. While less of a political hot-button today, the supposedly looming threat of massive food shortages was top of mind for many just a few decades ago. Paul Ehrlich is a long-time biology professor at Stamford University. In 1968, sponsored by the Sierra Club, Ehrlich published a best-selling book titled *The Population Bomb: Population Control or Race to Oblivion?* It began with the following apocalyptic prediction: "The battle to feed humanity is over. In the 1970s, hundreds of millions of people will starve to death in spite of any crash programs embarked upon now. At this late date, nothing can prevent a substantial increase in the world death rate."[84] On the other hand, Ehrlich would only concede that "[o]ther experts, more optimistic, think the ultimate food-population collusion will not occur until the decade of the 1980s." To

---

[84] P. Ehrlich, *The Population Bomb: Population Control or Race to Oblivion?* (The Sierra Club, 1968), p.1.

mitigate this catastrophe, Ehrlich proposed the immediate implementation of a variety of draconian population controls measures, "hopefully by a system of incentives and penalties, but by compulsion if voluntary methods fails."[85]

Ehrlich's dire predictions were based in a theatrical idol originating from Thomas Malthus's 1798 book, *An Essay on the Principle of Population*. This "Malthusian theory" posited that, while population had the tendency to grow exponentially, the world's ability to produce food would only grow linearly. Despite the widespread criticism of this theory, including for failing to account for innovation and the resiliency of food production, Malthusian models persisted into the Twentieth Century and undergirded Ehrlich's models and predictions. And because Ehrlich used these models to make such clear and near-term predictions, his prophesizing was very quickly proven false when hundreds of millions of people did not die of starvation in the 1970s. And even then, indeed decades later, Ehrlich attempted to partially stand by such predications. For example, Ehrlich told a reporter for the Earth Island Journal in 2009 that, in retrospect, he "feels that *The Population Bomb* was 'way too optimistic.'"[86] It seems the Idols of the Theater were truly tenacious for him.

While Ehrlich and *The Population Bomb* are a popular and easy target for this phenomenon, don't underestimate how widespread this feeling was in the scientific community at that time.

As another example, the New York Times published an article entitled *The End of Civilization Feared by Biochemist* on November 19, 1970. There, the Times reported that Harvard biochemist and Nobel Prize winner Dr. George Wald claimed that "civilization will end within 15 to 30 years unless immediate action is taken against problems facing man king," such as "pollution" and "overpopulation."[87] Environmental scientists told Life magazine in 1970 that "[i]n a decade, urban dwellers will have to wear gas masks to survive air pollution."[88] Many similar predications were made by the speakers at the first Earth Day celebration in 1970.

---

[85] Ehrlich, *The Population Bomb: Population Control or Race to Oblivion?*, p.1.
[86] T. Turner, *The Vindication of a Public Scholar*, Earth Island Journal, 2009, Accessed May 28, 2021, https://www.earthisland.org/journal/index.php/magazine/entry/the_vindication_of_a_public_scholar/
[87] New York Times, "The End of Civilization Feared by Biochemist," Nytimes.com, 1970, Accessed: May 28, 2021, https://www.nytimes.com/1970/11/19/archives/the-end-of-civilization-feared-by-biochemist.html
[88] H. Waters, "Why Didn't the First Earth Day's Predictions Come True? It's Complicated," Smithsonian Magazine, 2016, Accessed: May 28, 2021, https://www.smithsonianmag.com/science-nature/why-didnt-first-earth-days-predictions-come-true-its-complicated-180958820/

In the 1980s, "it was acid rain's turn to be the source of apocalyptic forecasts."[89] News magazine *Der Spiegel* in Germany reported that the forests were dying. Soil scientists predicted that a third of Germany's forests were already dead or dying and "[t]hey cannot be saved." When these disasters never came to pass, it turned out that the forests were "growing faster and healthier than ever" and that the rising acidity of lakes were "caused more by reforestation than by acid rain."[90]

In the 1990s, the watchword of climate change became global warming. A senior scientist at the Environmental Defense Fund said in 1990 that by 1996 the greenhouse gas effect "would be desolating the heartlands of North America and Eurasia with horrific drought, causing crop failures and food riots…a continent-wide black blizzard of prairie topsoil will stop traffic on interstates, strip paint from houses and shut down computers."[91] Similar to Ehrlich, this scientist recently affirmed, "[o]n the whole I would stand by these predictions."[92]

Environmental scientists also made incorrect predictions about global warming's impact on sea level. In 1988, it was predicted that "in the next 30 years" the island nation of the Maldives would be completely covered by the Indian Ocean.[93] In 2001, the Intergovernmental Panel on Climate Change updated that prediction, but again claimed that the Maldives are destined to disappear.[94] A minority of climate scientists dissented, warning that "[o]thers just want to repeat the same old flooding story," based on models or historical records rather than direct observation and experimentation. According to one Swedish scientist's model, global warming had caused increased evaporation in the ocean, resulting in the sea levels actually falling within the past few decades.[95]

---

[89] M. Ridley, "Apocalypse Not: Here's Why You Shouldn't Worry About End Times," Wired, 2012, Accessed: May 28, 2021, https://www.wired.com/2012/08/ff-apocalypsenot/
[90] Ibid.
[91] K. Smith, "Profits of doom," New York Post, May 29, 2011, Accessed May 28, 2021, https://nypost.com/2011/05/29/profits-of-doom/
[92] Ibid.
[93] M. Bastasch, "30 Years Ago Officials Predicted The Maldives Would Be Swallowed By The Sea. It Didn't Happen," The Global Warming Policy Forum, 2018, Accessed: May 28, 2021, https://www.thegwpf.com/30-years-ago-officials-predicted-the-maldives-would-be-swallowed-by-the-sea-it-didnt-happen/
[94] J. Pasotti, "Maldives Experience That Sinking Feeling," Science, 2005, Accessed: May, 28, 2021, https://www.sciencemag.org/news/2005/06/maldives-experience-sinking-feeling
[95] Ibid.

The 2000s saw a focus on melting of the ice caps and glaciers, fueled in no small part by Al Gore's Academy Award-winning *An Inconvenient Truth*. That 2006 film predicted that "within a decade, there will be no more snows of Kilimanjaro" and that the Arctic summers could be ice-free as soon as 2014.[96] Needless to say, you can still find snow on Kilimanjaro today. And a NASA study later found that increased Antarctic snow accumulation is more than offsetting the increased losses from thinning glaciers.[97] Again, none of this is to say that climate change is not occurring, or that humans are not a cause of it, or that we should not be engaging in efforts to mitigate or protect the environment. Rather, this history is important as providing examples of Idols of the Theater, where the People need to be on guard against catchy headlines based on artificial models can ultimately be harmful to the cause of science and the persuasion of others.

Another interesting Idol of the Theater exists in the various models that have been proposed for over a century on the extent, and possible exhaustion, of the Earth's oil reserves. Human nature has played an obvious role in these predictions. We all personally witness ourselves consuming and burning gallons after gallons, barrels after barrels, of gas and other fossil fuels day after day. It boggles the mind to think about the billions of other people engaged in similar activity, not just today, but for well over a century. Add to that the consumption we don't see but intuitively know exists—corporate consumption, electric power-plants, *etc.* Honestly, it seems like an utter miracle that the Earth has apparently supplied all these materials for so long. But we have been telling ourselves this story for a long, long time. It has become known as the perennial prediction of "peak oil."

In 1919, the chief geologist for the United States Geological Survey wrote that "the peak of production will soon be passed, possibly within 3 years," and that there were some "who present impressive evidence that it may come even before 1920."[98] In 1921, the president of the Colorado School of Mines wrote that "[t]he average middle-aged man of today will live to see the virtual

---

[96] B. Lomborg, "Al Gore's Climate Sequel Misses a Few Inconvenient Facts," *Wall Street Journal*, July, 27, 2017, Accessed: May 28, 2021, https://www.wsj.com/articles/al-gores-climate-sequel-misses-a-few-inconvenient-facts-1501193349

[97] M. Vinas, "Mass gains of Antarctic Ice Sheet greater than losses, NASA study reports," Phys.org., 2015, Accessed May 28, 2021, https://phys.org/news/2015-10-mass-gains-antarctic-ice-sheet.html

[98] D. White, "The unmined supply of petroleum in the United States," *Transactions of the Society of Automotive Engineers* 14, 227 (1919).

exhaustion of the world's supply of oil from wells."[99] In 1937, the director of the United States' naval petroleum reserves told Congress that "[t]he best information is that the present [oil] supply will last only 15 years. That is a conservative estimate."[100]

In 1956, Marion King Hubbert created a well-known and long venerated model for oil production, which assumed that oil production would follow a symmetric, bell-shaped curve.[101] This "Hubbert Curve" predicted that peak oil production would be reached when approximately half of the world's oil resources are exhausted. Hubbert predicted that U.S. production would peak in 1970 and for the whole world by 2000. Hubbert's predictions were updated in 1998 by Colin Campbell and Jean Laherrere in a paper published in Scientific American, this time predicting peak oil in 2004-2005, which would then "start an irreversible decline."[102] Together, they founded the "Association for the Study of Peak Oil and Gas" (ASPO), which published a popular newsletter. Before it was discontinued in 2009, the newsletter updated their 1998 prediction a few years later to predict peak oil in 2008. Of course, none of these predictions over the past hundred-plus years has been close to correct.

Hubbert's peak oil model, and the peak-oil-idol in general, began to die down in the most recent decade. As energy economist Michael Lynch wrote in *What Ever Happened to Peak Oil?*:

> Unfortunately, very few people realize[d] that the entire concerns about peak oil were based on misinformation or junk science. Specifically, the research was not scientific at all but statistical analysis so badly done that it wouldn't pass a first-year college course. The work by Campbell and Laherrere relied on the basic idea that geology determined production trends…The majority of this is nonsense. Production usually doesn't follow a bell curve, and when it does, it is the result of the effects of exponential growth and decline….Instead, changes in oil prices, fiscal terms, and access to resource basins cause production to fluctuate all the time—and often surpass the supposed 'peak' level that peak oil advocates identify. Many of the arguments reflected their authors' ignorance of either the industry or forecasting… Thus, the

---

[99] V. Anderson, "The present status of the oil shale industry," *Colorado School of Mines Quarterly* 16, 12 (1921).
[100] San Bernardino Sun, "Senate Committee Given Report on Threatened Petroleum Shortage," San Bernardino Sun, Mar. 10, 1937.
[101] U. Bardi, "Peak Oil, 20 years later: Failed prediction or useful insight." *Energy Research & Social Science* 48, 257 (2019).
[102] Ibid.

publications and predictions have by and large not come true—often rather spectacularly.[103]

Though of course, it should come as no surprise that the covid-19 pandemic has renewed claims that peak oil is "suddenly upon us."[104]

Idols of the Theater originate in the human inclination to want to simplify the natural world so that it is understandable, predictable, and accords with the narratives (philosophical, political and moral) that we want to tell ourselves. Scientists want to have the answers. As the quote at the beginning of this chapter suggests, they want to "discern Things in their Causes" and "trace the ways of highest Agents." So they build artificial models that are simple and intelligible and that are supposed to represent the world. Sometimes they work, but often they do not. The real world is messy and complicated. Nature is something counterintuitive and weird. The People have a role to play when a science becomes overly beholden to a single model that fails to live up to its predictions. They have a role in preventing science from becoming dogmatic, angry, and overly-defensive about criticism. Science must always be open to the beauty and complexity of nature in itself.

Conclusion

In his 1784 essay entitled *Answering the Question: What is Enlightenment?*, Immanuel Kant suggested that the motto for the Enlightenment should be *sapere aude*—"dare to know." Is that motto in tension with the modern rejoinder, "trust the science," at least when interpreted as demanding unquestioning, quasi-blind trust from lay people?

The People have a responsibility to think freely and independently about science, rejecting both rigid dogmatism and over-confidence as well as unruly and unfair skepticism and naysaying. Like all scientists, we are humans and can recognize Idols of the Tribe. We can tell when science is bending over backwards to be as candid as possible, and when shady methodology does not pass the smell test. Like all scientists, we know that people can have agendas, prejudices, worldviews, or other Idols of the Cave that will shade their conclusions. We can also be alert to statistical shenanigans and other imprecise Idols of the Marketplace. Finally, we should not be reactionary to theatrical,

---

[103] M. Lynch, "What Ever Happened To Peak Oil?," *Forbes*, June 29, 2018, Accessed: May 28, 2021, https://www.forbes.com/sites/michaellynch/2018/06/29/what-ever-happened-to-peak-oil/?sh=12406fdf731a

[104] T. Randall and H. Warren, "Peak Oil Era Is Suddenly Upon Us," Bloomberg.com, Dec. 1, 2020, Accessed May 28, 2021, https://www.bloomberg.com/graphics/2020-peak-oil-era-is-suddenly-upon-us/

headline-grabbing doomsayers demanding immediate existential changes to society…or else. Such things require sober, empirical analysis that meets the world as it truly is, rather than in a cramped, simplified, neatly ordered, convenient, and artificial Idol of the Theater. The temptations that Bacon identified centuries ago as calling scientists to stray from the straight and narrow path have always been there and will continue to exist as long as we are fallible humans.

Thus, to summarize, we have seen how America's founders engineered the Constitution to allow the People to rule themselves while preventing the tyranny of a bare majority. They did this by turning a perceived weakness of democracy (faction and pursuit of individual interest) into a strength that could be made to actually serve the interests of Liberty. The rise of the administrative state under the Progressivism of the early Twentieth Century was not only inherently inconsistent with these founding principles, it was openly hostile to them in preference to rule by efficient experts. But this hope for a depoliticized politics, for a scientific morality, for a utopian Rationalia, is philosophically impossible. Even if scientists could know perfectly what "is," nothing can give them a privileged insight for policy into what "ought to be." The People must be brought to bear to make value judgments, assigning the weight such values warrant for those scientific findings. And now, we have seen that the People also have a significant role to play as collaborators and helpers of science, as outsiders capable of applying common sense and their own wealth of experience on human nature to scientific theories and assertions.

The final factor to consider is what moral influence political power might have on the expert bureaucrat, and whether its potential for corruption and dehumanization requires such power to be diffused as much as possible among the People as a whole, rather than vesting in the expert.

6.
Ideology Fills the Vacuum

I can see the . . . seams where they're put together. And, almost, see the apparatus inside them take the words I just said and try to fit the words in here and there, this place and that, and when they find the words don't have any place ready-made where they'll fit, the machinery disposes of the words like they weren't even spoken.

—Chief Bromden, *One Flew over the Cuckoo's Nest* by Ken Kesey

Oliver Wendell Holmes, Jr.

We all know the stories and likely have many of our own: petty bureaucrats with Napoleon complexes and labyrinthine administrative procedures filled with Catch-22's and inhuman nightmares out of Kafka. We intuitively know the danger of even a little fiefdom of power in the hands of a person apt to let it go to his head. But these cliché situations are not the point here. The issue we are addressing is more complicated, and more fundamental. The real question is what happens when political power is exercised through someone who believes that such power should be scientifically and efficiently deployed because it does not derive solely from the consent of the governed. What happens to the one who exerts such power? And what happens to those it is exerted upon? Rather than appealing to ubiquitous anecdotes about the DMV, the case of Supreme Court Justice Oliver Wendell Holmes, Jr. will provide the more illuminative example of this more fundamental objection to power vesting in the hands of the expert few.

Holmes grew up in an aristocratic Bostonian family with strong connections to the most prominent intellectual circles. While attending Harvard, he interacted with Ralph Waldo Emerson and became engaged by the prevailing American transcendentalist and pragmatist philosophies of the time. After fighting in the Civil War, Holmes attended Harvard Law School, began a legal practice, mingled among the high society of Boston and London, and developed a distinguished academic reputation in the law before being appointed to the United States Supreme Court by Theodore Roosevelt.

Holmes remains a titan of American law and jurisprudence. He was an admirable and highly talented judge in many ways. After two communists were prosecuted in New York under Woodrow Wilson's Sedition Act for distributing leaflets condemning American interference with Russia's communist revolution, Holmes dissented when the majority of the Supreme Court upheld their conviction.[1] Holmes explained that "we should be eternally vigilant against attempts to check the expression of opinions that we loathe and believe to be fraught with death, unless they so imminently threaten immediate interference with the lawful and pressing purposes of the law that an immediate check is required to save the country."[2] He maintained unlikely restraint and humility, endorsing the "marketplace of ideas" concept of free speech:

> If you have no doubt of your premises or your power, and want a certain result with all your heart, you naturally express your wishes in law, and sweep away all opposition...But when men have realized that time has upset many fighting faiths, they may come to believe even more than they believe the very foundations of their own conduct that the ultimate good desired is better reached by free trade in ideas -- that the best test of truth is the power of the thought to get itself accepted in the competition of the market, and that truth is the only ground upon which their wishes safely can be carried out. That, at any rate, is the theory of our Constitution.[3]

In another case, it was Holmes that devised the well-known "clear and present danger" test for free speech and the perennially popular "shouting fire in a crowded theater" example.[4]

As admirable and influential as his opinions in these cases at the Supreme Court were, Holmes was ultimately not driven by the same underlying philosophies of the Founding Fathers. He did not believe that natural law or natural rights existed, much less ones that endowed all people with inalienable liberties pre-existing government. He did not agree that the goal of government is to be the protector and securer of Liberty. Holmes, like other progressives, was a legal positivist. Legal positivism contends that the law is, and always has been, simply what human beings with the relevant power said it was. Whatever was posited as the law was the law.

This should sound familiar because it is the same underlying philosophy as Maximilian Robespierre and Woodrow Wilson. It is the idea that, whatever liberties you may have, they come from the government. You get to enjoy them

---

[1] *Abrams v. United States*, 250 U.S. 616 (1919) (Holmes, J. *dissenting*).
[2] Ibid.
[3] *Abrams v. United States*, 250 U.S. 616 (1919) (Holmes, J. *dissenting*).
[4] *Schenck v. United States*, 249 U.S. 47 (1919).

at the pleasure and will of the government, and they exist because they are beneficial for the purposes of the collective constituting a society.

Along with Sir William Blackstone, the quintessential English Enlightenment philosopher of law was Sir Edward Coke, who wrote in 1628 that "[r]eason is the life of the law, nay the common law itself is nothing else but reason."[5] In his early and most famous academic work, Holmes consciously turned Coke's observation on its head, beginning his treatise on *The Common Law* with the claim that "[t]he life of the law has not been logic; it has been experience."[6] There is no law of nature. The law arises from the "felt necessities of the time, the prevalent moral and political theories, intuitions of public policy, avowed or unconscious, even the prejudices which judges share with their fellow-men." According to Holmes, the law evolves with "the story of a nation's development" and its substance "at any given time pretty nearly corresponds, so far as it goes, with what is then understood to be convenient."[7] These are the basic tenets of legal positivism.

Moreover, Holmes did not hesitate to criticize natural law and natural rights. As he later wrote as a Supreme Court Justice, "[t]he common law is not a brooding omnipresence in the sky, but the articulate voice of some sovereign or *quasi*-sovereign that can be identified."[8] Just as human societies evolve and develop, so too must its law. Thus, Holmes lent an early voice to what seem ubiquitous today: seeing the Constitution as a living document and the law as evolving rather than eternal and natural. In other words, Holmes' scorn for a "brooding omnipresence" of law was clearly at odds with the Enlightenment views of the founders, who predicated the Declaration of Independence on "the Laws of Nature and of Nature's God." Accordingly, in another case, Holmes explained that "[t]he provisions of the Constitution are not mathematical formulas that have their essence in form, they are organic, living institutions transplanted from English soil. Their significance is vital, not formal."[9]

For Holmes as with all others who reject natural law, another question must arise once the natural law is swept aside as an old-fashioned fantasy. If we must make the law for ourselves because it just is whatever we posit it to be, how should it be made? And it should come as no surprise that the answer Holmes espoused was science. In charting *The Path of the Law* in 1897, Holmes warned:

---

[5] E. Coke, *The First Part of the Institutes of the Laws of England* (1628), bk. 2, ch. 6, section 138.
[6] O.W. Holmes, Jr., *The Common Law* (1881).
[7] Ibid.
[8] *Southern Pacific Co. v. Jensen*, 244 U.S. 205 (1917).
[9] *Gompers v. United States*, 233 U.S. 604 (1914).

> We must beware of the pitfall of antiquarianism, and must remember that for our purposes our only interest in the past is for the light it throws upon the present. I look forward to a time when the part played by history in the explanation of dogma shall be very small, and instead of ingenious research we shall spend our energy on a study of the ends sought to be attained and the reasons for desiring them. As a step toward that ideal it seems to me that every lawyer ought to seek an understanding of economics.[10]

In other words, "[f]or the rational study of the law...the man of the future is the man of statistics and the master of economics."[11] Accordingly, like most progressives, Holmes found it just and proper for the law to "emphasize the criterion of social welfare as against the individualistic eighteenth-century bill of rights."[12]

But if Holmes was just another social-welfare progressive that deferred little, if at all, to the Enlightenment philosophy underlying the Constitution and the Declaration of Independence, his would not be a particularly interesting case. What makes Holmes interesting is that his legal positivism freed him to follow the science of his day beyond mere socialism of property to its logical conclusion of placing the progress of the collective over the individual *qua* human. Like Woodrow Wilson, Holmes embraced the prevailing scientific views of eugenics, writing:

> I believe that the wholesale social regeneration which so many now seem to expect, if it can be helped by conscious, coordinated human effort, cannot be affected appreciably by tinkering with the institution of property, but only by taking in hand life and trying to build a race. That would be my starting point for an ideal for the law. The notion that with socialized property we should have women free and a piano for everybody seems to me an empty humbug.[13]

And for a legal positivist, if science suggested we can improve the human race scientifically, why shouldn't the law lend its hand to help? As Holmes would tell a friend, "eugenics...was the true beginning, theoretically, of all improvement. The folly, to my mind of socialism is that it begins with property instead of with life."[14]

---

[10] O.W. Holmes, Jr., "The Path of the Law," *Harvard Law Review* 10, 457 (1897).
[11] Ibid.
[12] O.W. Holmes, Jr., "Ideals and Doubts," *Illinois Law Review* 10, 1 (1915).
[13] O.W. Holmes, Jr., "Ideals and Doubts," *Illinois Law Review* 10, 1 (1915).
[14] O.W. Holmes, Jr., Letter to Lady Leslie Scott, May 17, 1912.

It was this, and only this, extension of scientific political power that has made modern progressives disown Holmes. Holmes believed every bit as much in the law being allowed and encouraged to serve the interests of the collective. He just happened to think the science suggested that the master race was the way to go. While fighting in the Civil War, Holmes wrote a letter to his sister saying that, while he was a democrat in theory, when he was among the *hoi polloi* he found that he "loathe[d] the thick-fingered clowns we call the people."[15] Confessing that he was "living *en aristocrat*," Holmes believed that there were "only two civilized placed in America," Boston and Philadelphia.[16]

Now that the supposed science of eugenics has been debunked, modern progressives view Holmes as an "aristocratic nihilist" and "a cold and brutally cynical man who had contempt for the masses and for the progressive laws he voted to uphold."[17] But since basically no one in America today believes in eugenics, Holmes' example provides unique insight to our modern eyes into how political power operates within someone with a legal positivist view guided by science. What does such power do when it runs up against the individual Liberty of a human being? And for that purpose, there is one other person whose life is worth comparing to that of Oliver Wendell Holmes, Jr.

Carrie Elizabeth Buck

In 1906, Carrie Buck was born in rural Virginia. Her father either left or died when she was a baby, and her mother, Emma Buck, reportedly became a prostitute.[18] She was placed with foster parents, the Dobbs, when she was three years old, and removed from school at twelve to help the Dobbs with household tasks. At seventeen, Buck claimed to have been raped and impregnated by the Dobbs' nephew. Within a few months, in 1924, the Dobbs had Buck was committed to the Virginia Colony for the Epileptic and Feebleminded because she was allegedly "feeble minded"—a eugenics watchword essentially meaning undesirable…and/or promiscuous. Buck's daughter, Vivian, was born there three months later. Vivian would die from measles at the age of six.

In 1924, Virginia's General Assembly had passed its "Eugenical Sterilization Act," which allowed the superintendent of the Virginia Colony, Albert Priddy, to sterilize those under his care. Priddy immediately commenced proceedings to have Buck sterilized by cutting her fallopian tubes. Irving Whitehead, a known

---

[15] O.W. Holmes, Jr., Letter to Emily Hallowell, November 16, 1862.
[16] Ibid.
[17] J. Rosen, "Brandeis's Seat, Kagan's Responsibility," New York Times, July 3, 2010, Accessed: May 28, 2021, https://www.nytimes.com/2010/07/04/opinion/04rosen.html
[18] *See generally* P. Thompson, "Silent Protest: A Catholic Justice Dissents in Buck v. Bell," *The Catholic Lawyer* 43, 125 (2004).

eugenicist and close friend of Priddy, served as Buck's "attorney"—if one could call him that—for the sterilization proceedings. The case made its way to the United States Supreme Court, which rendered an 8 to 1 decision on May 2, 1927, allowing Buck's sterilization to proceed at the hands of the Colony's surgeon, John Bell. A motion asking the Court to reconsider was denied in October, and Dr. Bell plied his "eugenical" trade on her on October 19, 1927.

After suffering this irreversible sterilization, Carrie Buck lived into the 1980s and was noted for being "an avid reader and a lucid conversationalist, even in her last days."[19] "In 1982, [Buck], who had wanted children all her life, played the role of the Virgin Mary in a Christmas play. She died a few weeks later and was buried in Lynchburg only a few steps from her daughter Vivian...[Before her death,] she forgave those who had treated her so badly and declared, 'I tried helping everybody all my life, and I tried to be good to everybody. It just don't do no good to hold grudges.'"[20]

It was Oliver Wendell Holmes, Jr. who wrote the Court's majority opinion in the case of Buck v. Bell, giving Dr. John Bell the permission of the law to permanently sterilize Carrie Buck. The Court's 1927 opinion is important, not so much for what it says about eugenics, which has since fallen out of favor as a "science," but for what it shows about how those with political power, operating under a collectivist and "scientific" legal philosophy, see individuals.

Holmes' perspective on people like Buck comes through clearly in the language of his opinion. He summarized the facts of the case succinctly as follows: "Carrie Buck is a feeble-minded white woman who was committed to the State Colony above mentioned in due form. She is the daughter of a feeble-minded mother in the same institution, and the mother of an illegitimate feeble-minded child."[21] And he equally succinctly explained the reason why he thought the law could call upon Buck to forfeit her human reproductive ability:

> We have seen more than once that the public welfare may call upon the best citizens for their lives. It would be strange if it could not call upon those who already sap the strength of the State for these lesser sacrifices, often not felt to be such by those concerned, in order to prevent our being swamped with incompetence. It is better for all the world if, instead of waiting to execute degenerate offspring for crime or to let them starve for their imbecility, society can prevent those who are manifestly unfit from continuing their kind. The principle that sustains

---

[19] Thompson, "Silent Protest: A Catholic Justice Dissents in Buck v. Bell."
[20] Thompson, "Silent Protest: A Catholic Justice Dissents in Buck v. Bell."
[21] *Buck v. Bell*, 274 U.S. 200 (1927).

compulsory vaccination is broad enough to cover cutting the Fallopian tubes...Three generations of imbeciles are enough.[22]

It is truly hard to believe that such words are found in the 1927 reports of the United States Supreme Court, and not in some German official's edicts ten years later.

It is also surprising that the usually reserved Oliver Wendell Holmes, Jr. seemed to almost relish what he had accomplished through the case of Buck versus Bell. The seven other Supreme Court Justices that joined Holmes' opinion in the case had initially balked at the way Holmes wrote the opinion, and suggested ways to soften the language. Holmes, however, confided in his friend, Harold Laski, "I am amused (between ourselves) at some of the rhetorical changes suggested, when I purposely used short and rather brutal words."[23] Indeed, he would later reminisce in other correspondence that "[o]ne decision that I wrote gave me pleasure, establishing the constitutionality of a law permitting the sterilization of imbeciles."[24]

Just about everyone today knows that eugenics is wrong after we saw the horrors of World War II, which Holmes did not have the benefit of to reevaluate his views. But the point here is that once the natural law is abandoned, something must come and fill its void. Science itself, and its legal equivalent found in Holmes' positivism, may provide a framework, but it has no substance, no mechanism of weighing is's to find ought's. The vacuum left by rejection of natural law requires an ideology to fill it. Holmes confided to a friend that he did not think "nobly of man."[25] He "look[ed] at men...as like flies—here swept away by a pestilence—there multiplying unduly and paying for it."[26] In such a world, the natural law or supposed morals were "only a check for varying intensity on force," which was the real "ultimate."[27] Instead, Holmes' ideology was based on a sort of Nietzschean view of strength towards an ideal *Übermensch*.[28] In one of the most famous speeches Holmes ever delivered— *The Soldier's Faith*—Holmes affirmed that "[t]he joy of life...is to put out all one's powers as far as they will go; that the measure of power is obstacles overcome; to ride boldly at what is in front of you, be it fence or enemy; to pray,

---

[22] *Buck v. Bell*, 274 U.S. 200 (1927).
[23] O.W. Holmes, Jr., Letter to Harold Laski, April 29, 1927.
[24] O.W. Holmes, Jr., Letter to Lewis Einstein, May 19, 1927.
[25] O.W. Holmes, Jr., Letter to Harold Laski, July 23, 1925.
[26] O.W. Holmes, Jr., Letter to Harold Laski, July 23, 1925.
[27] O.W. Holmes, Jr., Letter to Harold Laski, July 23, 1925.
[28] S. Vannatta and A. Mendenhall, "The American Nietzsche? Fate and Power in the Pragmatism of Justice Holmes," *UMKC Law Rev.* 85, 187 (2017).

not for comfort, but for combat."²⁹ The science of eugenics was an easy and obvious fit for this ideology.

When political power indulges the efficiency of science mixed and mingled with the coldness of an ideology, the result is a kind of blindness that explains not only the result of Buck versus Bell, but also the ruling's obviously harsh language and the apparent relish Holmes took in delivering it up. Holmes could not see the humanity of Carrie Buck because his ideology sought things beyond the human. It sought a goal, an ideal, for the collective of society.

Only the natural law codified in America's founding documents keeps the eyes of political power open to humanity in itself. The natural law cares only to defend, support, and serve the pre-existing rights and liberties of the People. Natural law is not an ideology. It is a light shined upon humanity itself. On the other hand, political power exercised scientifically to achieve an adopted goal inevitably blinds those who exercise it to the humanity of those they purport to serve.

Pierce Butler

Before considering other examples of this phenomenon beyond Carrie Buck, a final coda is found in the lone dissent in the Supreme Court from Justice Pierce Butler in her case. While Holmes is ranked as one of the most influential justices ever to serve on the Court, the memory of Butler today is not as widespread.[30] Butler was born in a log cabin in rural Minnesota in 1866 to Irish parents who had fled the potato famine of 1848. Butler and his eight siblings lived "a humble life as frontier farmers" and attended a one-room schoolhouse. Butler learned law as an apprentice in St. Paul, and earned a good reputation as a smart, tough, but fair prosecutor before being appointed to the Supreme Court.

Butler's opinions on the Court demonstrated his keen sense of individual liberty, the natural law of morality, and natural rights. He read the Constitution to forbid the state to "take from the individual the right to engage in common occupations of life."[31] Butler's opposition to New Deal regulations earned him a spot as one of the so-called "Four Horsemen," who consistently voted to strike down such laws. Butler also dissented in a case where the Court upheld wiretapping without a warrant, saying that it was important to "construe[] the Constitution in light of the principles upon which it was founded."[32] (In the

---

[29] O.W. Holmes, Jr., *The Soldier's Faith*. (1895).
[30] Thompson, "Silent Protest: A Catholic Justice Dissents in Buck v. Bell."
[31] *Senn v. Tile Layers Protective Union*, 301 U.S. 468 (1937).
[32] *Olmstead v. United States*, 277 U.S. 438 (1928).

1960s, the Supreme Court finally changed course on the wiretapping issue, agreeing with Butler.) Rights and liberty were always placed at the forefront by Butler. Even in cases involving criminals, he affirmed that "[a]bhorrence, however great, of persistent and menacing crime will not excuse transgression in the courts of the legal rights of the worst offenders."[33]

Natural law and the founding principles of the country were Butler's guiding light in exercising the political power he was afforded. While the Supreme Court was deliberating on the Buck v. Bell decision, Holmes quickly won the approval of the other seven Justices. During this process, Holmes was quoted to remark to another justice, "I bet you Butler is struggling with his conscience…He knows the law is the way I have written it. But he is afraid of the Church. I'll lay you a bet that the Church beats the law."[34] But Butler was indefatigable, and Holmes later quipped that Butler was a "monolith, there are no seams the frost can get through."[35] Nothing could blind him to the humanity of Carrie Buck.

The underlying legal philosophies of Holmes and Butler simply cannot be reconciled. Holmes' emphasis on science, collectivism, and legal positivism are all of a piece, just as Butler's emphasis on natural law, individual rights, and universal reason was for him. They resulted in fundamentally different and incommensurate ways of looking at Carrie Buck. For one Justice, she was a means to an end; for the other, she was a human being, whose liberty was the one and only end to be sought by the servants of her government.

The Humanitarian Theory of Punishment

Science today does not support eugenics. But science is still brought to bear on a variety of subjects that touch upon our humanity, and no other political arena has felt its influence as much as the criminal justice system. Unlike eugenics, the modern application of science in this field claims to be milder and more "humanitarian" than the natural law. On the contrary, modern science now views natural law itself as vulgar and barbarous. But what vision of humanity does the so-called "humanitarian" theory of punishment actually endorse?

Since the earliest forms of law like Hammurabi's Code and the Law of Moses, criminal justice had been organized around the concept of "crime and punishment." This traditional, and now much-maligned, account has become known today as the retributive theory of justice. Often, the theory of retribution

---

[33] *United States v. Motlow*, 10 F.2d 657 (7th Cir. 1926).
[34] M. Urofksy, *Biographical Encyclopedia of the Supreme Court: The Lives and Legal Philosophies of the Justices* (CQ Press, 2006), p. 107-108.
[35] Urofksy, *Biographical Encyclopedia of the Supreme Court: The Lives and Legal Philosophies of the Justices*, p. 107-108

it espouses is distorted. Some say retributive justice is simply an emotional response to even things out, "an eye for an eye." Others contend that it serves as a stand-in for what would otherwise be vigilante justice or vengeance by the victim or the victim's family. (Recall here Sam Harris's contempt for the Albanian tradition of vendetta known as *Kanun*.)

Retributive justice is actually none of those things. For example, while the Law of Moses spoke about the justice of "an eye for an eye," it also reports God clarifying that "vengeance is mine."[36] Retributive justice is only the idea that it is a person's *choice* to commit crime justifies punishment. The imprisonment, the fine, or the other sanction is deserved because the natural law has been broken. Justice is not about the victim or the criminal; it is about the law which has been broken.

Alexander Solzhenitsyn knew the perils of a warping the system of criminal justice away from the retributive theory better than most after spending years in the Soviet gulags. He recognized that all justice finds its basis in that which pre-exists government and is common to humanity. He wrote, "[j]ustice is conscience, not a personal conscience but the conscience of the whole of humanity. Those who clearly recognize the voice of their own conscience usually recognize also the voice of justice."[37]

This retributive theory of justice, based on natural law, prevailed through the Age of Enlightenment. Most famously, Immanuel Kant argued that retributive justice is the only form of punishment a court can properly render. The criminal law, according to Kant, "can never be used merely as a means to promote some other good for the criminal himself or for civil society."[38] Punishment must only be imposed based on the crime itself because otherwise the justice system would be manipulating a human being "to the purposes of someone else" and being treated as an "object." The criminal's "innate Personality (that is, his right as a Person) protects him against such treatment."[39] As the ubiquitous English legal commentator, Sir William Blackstone, summarized: the law had always accepted that "punishments are…only inflicted for abuse of that free-will which God has given to man."[40]

---

[36] Deuteronomy 32:35.
[37] L. Labedz and A. Solzhenitsyn, *Solzhenitsyn* (New York: Harper & Row, 1971), translating Solzhenitsyn Letter to Students, Oct. 1967.
[38] I. Kant, *Metaphysical Elements of Justice* (Hackett Publishing, 1990), p. 138.
[39] Kant, *Metaphysical Elements of Justice*, p. 138.
[40] W. Blackstone and R. Kerr, *Commentaries on the laws of England* (London: J. Murray, 1880), vol. 4, section 27.

However, other theories of criminal justice rose to prominence around the same time that Woodrow Wilson's ideas on the efficient administration of government did. This is no coincidence. Both shared the same philosophic basis in (i) a rejection of natural law and the founding principles of liberty, (ii) an emphasis on the collective over the individual human, and (iii) the employment of science to efficiently achieve collective goals. In 1909, for example, the American legal scholar Roscoe Pound reviewed and touted the "scientific" work being done in the fields of "the Science of Criminology," "Criminal Anthropology," and "Sociology."[41] He lamented that these new sciences had not be brought to bear on the American criminal justice system. "[C]riminal law is the most archaic part of our legal system" because it "is so rooted in theological ideas of free will and moral responsibility and juridical ideas of retribution...that we by no means make what we should of our [scientific] discoveries." He predicted that criminal law "need and will soon receive" this "very different form of study" based on "what continental scientists and philosophers have been doing and thinking."[42]

Pound was right. Scientists operating in the fields touted by Pound began to claim they had discovered that people were nothing more than products of their biology, their upbringing, and their culture. Free will did not exist in a deterministic world. Accordingly, as one American psychologist put it, "[t]he time seems to have come when psychology must discard all reference to consciousness; when it need no longer delude itself" that mental states actually exist.[43] Accordingly, Pound criticized the American legal system's "exaggerated respect for the individual" based on science having "routed" the concept of the individual and a person's free will."[44] Progressive reformers called instead for the criminal law to abandon the retributive theory of justice in favor of therapeutic treatment (reformation) and deterrence.

Oliver Wendell Holmes approved of the deterrence theory. In his 1881 treatise on *The Common Law*, he wrote that "[p]revention would accordingly seem to be the chief and only universal purpose of punishment. The law threatens certain pains if you do certain things...If you persist in doing them, it has to inflict the pains in order that its threats may continued to be believed."[45]

---

[41] R. Pound, "Book Review," *American Political Science Review* 3, 281 (1909).
[42] Ibid.
[43] J. Watson, "Psychology as the Behaviorist Views It," *Psychological Review* 20, 158 (1913).
[44] A. Alschuler, "The Changing Purposes of Criminal Punishment: A Retrospective on the Law Century and Some Thoughts about the Next," *University of Chicago Law Review* 70, 1 (2003).
[45] O.W. Holmes, Jr., *The Common Law* (1881).

Holmes denied the existence of free will. And so he proposed that, if he was talking to someone that was going to be hanged, he would say: "I don't doubt that your act was inevitable for you but to make it more avoidable by others we propose to sacrifice you to the common good."[46]

Other progressives preferred the reformation theory. Progressive reformer Gino Carlo Speranza wrote (mixing in some eugenics for good measure):

> The conception of punishment as a defence to crime has gone into bankruptcy: it neither defends nor deters. Criminal therapeutics must take its place; that is, where a cure is possible, let the remedial agencies suggested by criminologic and sociologic science have full scope. But where juridic therapeutics fails, let there be no mistaken altruism to perpetuate the unfittest.[47]

The retributive theory fell by the wayside for those that espoused deterrence and those that preferred reformation alike because neither recognized a pre-existing liberty interest based on the humanity of the individual. As John Wigmore explained, "[t]he measures of the modern penal law are not based on moral blame."[48] He likened criminals to weeds in a garden—we can pull the "human weed" even if it was "predetermined by nature and environment to do just what he did."[49]

These new views on the nature of criminal justice (or better yet, criminal medicine or criminal gardening) had tangible impacts on its administration. Penal codes began to emphasize the need for indeterminate sentences and treatment for criminals. For example, the Model Penal Code issued by the American Law Institute in 1962 required that any sentence for a felony needed to be for at least one year so that correctional authorities could have sufficient time to diagnose the offender. Scientific committees were formed to oversee these sentences—parole boards, probation officers, dependency counsellors, psychologists and psychiatrists were all enlisted to help. Those with the deterrent view would say that the criminal ought to be locked up until the scientist declares him to be no longer dangerous. Those with the reformation

---

[46] O.W. Holmes, Jr., Letter to Harold Laski, December 17, 1925.
[47] Alschuler, "The Changing Purposes of Criminal Punishment: A Retrospective on the Law Century and Some Thoughts about the Next."
[48] Alschuler, "The Changing Purposes of Criminal Punishment: A Retrospective on the Law Century and Some Thoughts about the Next."
[49] Alschuler, "The Changing Purposes of Criminal Punishment: A Retrospective on the Law Century and Some Thoughts about the Next."

view would wait until the scientist declares, "he's cured!" Neither cared about what the offender deserved.

But against this tidal wave washing away the retributive theory of justice, C.S. Lewis provided an able defense of the traditional theory in his essay, *The Humanitarian Theory of Punishment.* There, he explained why he thought that those who ascribed to the humanitarian theories based on a belief that they were "mild and merciful" were "seriously mistaken."[50] For such people, "it appears at first sight that we have passed from the harsh and self-righteous notion of giving the wicked their deserts to the charitable and enlightened one of tending the psychologically sick. What could be more amiable?" But, according to Lewis, the so-called "humanity" of the deterrence and reformation theories is "a dangerous illusion and disguises the possibility of cruelty and injustice without end."[51] Lewis advanced three arguments to prove this.

First, when the criminal justice system thinks only of deterrence and reformation, it has necessarily abandoned its claim to justice. "There is no sense in talking about a 'just deterrent' or a 'just cure.' We demand of a deterrent not whether it is just but whether it will deter. We demand of a cure not whether it is just but whether it succeeds."[52] These theories inherently remove the criminal from the sphere of justice, from being treated as a human being with "rights," to the sphere of science to be studied as an object or "case." Moreover, Lewis foresaw that once punishment is untethered from justice, crime inevitably would be as well. "For if crime and disease are to be regarded as the same thing, it follows that any state of mind which our [experts] choose to call 'disease' can be treated as a crime; and compulsorily cured."[53] Why can't thoughts be just as much a crime as murder if criminal law is about protecting society and curing the criminal? Crime can now be defined ideologically rather than by the traditional light of the natural law.

Second, since the issue is not about what is "just," judges and juries have little to no role to play in administering criminal justice. "Only the expert 'penologist' (let barbarous things have barbarous names), in the light of previous experiment, can tell us what is likely to deter: only the psychotherapist can tell us what is likely to cure."[54] The People are the arbiters of what is morally right,

---

[50] C. S. Lewis, "Humanitarian Theory of Punishment by C.S. Lewis," Matiane, 1953, Accessed: May 28, 2021, https://matiane.wordpress.com/2018/10/24/humanitarian-theory-of-punishment-by-c-s-lewis/
[51] Lewis, "Humanitarian Theory of Punishment by C.S. Lewis."
[52] Lewis, "Humanitarian Theory of Punishment by C.S. Lewis."
[53] Lewis, "Humanitarian Theory of Punishment by C.S. Lewis."
[54] Lewis, "Humanitarian Theory of Punishment by C.S. Lewis."

but they are not "experts" in the fields that the deterrence and reformation theories require. The People are therefore forced to entrust the criminal into the hands of "technical experts whose special sciences do not even employ such categories as rights or justice." And such experts "can be criticized only by fellow-experts and on technical grounds, never by men as men and on grounds of justice."[55]

Third, once the administration of "criminal justice" is entrusted to the expert, there is every reason to believe that its outcome will be cruel and unjust under the traditional view, albeit unintentionally so.

> Of all tyrannies, a tyranny sincerely exercised for the good of its victims may be the most oppressive. It may be better to live under robber barons than under omnipotent moral busybodies. The robber baron's cruelty may sometimes sleep, his cupidity may at some point be satiated; but those who torment us for our own good will torment us without end for they do so with the approval of their own conscience.[56]

Under the retributive theory, once a criminal has served his term, he is released because he has "paid his debt." But the scientist, who rejects ideas of "just deserts," will have no qualms about keeping a criminal under his care until he is no longer a danger to society and/or no longer sick. To let the criminal loose prematurely would just be failure. And yet it would be useless to claim that a "punishment" was hideously disproportionate to the crime because:

> The experts with perfect logic will reply, "but nobody was talking about deserts. No one was talking about punishment in your archaic vindictive sense of the word. Here are the statistics proving that this treatment deters. Here are the statistics proving that this other treatment cures. What is your trouble?"[57]

There is no room for mercy under the theories of deterrence and reformation. If crime is a disease and not a choice based on free-will, "[h]ow can you pardon a man for having a gumboil or a club foot?"[58]

The criminal justice system and the debate between the retributive theory founded in natural law and the deterrence and reformation theories founded in science are, of course, only one example of the underlying issue for the purposes here. The point is to show what necessarily happens when science is no longer subservient to, but rather displaces, the natural law. The scientist

---

[55] Lewis, "Humanitarian Theory of Punishment by C.S. Lewis."
[56] Lewis, "Humanitarian Theory of Punishment by C.S. Lewis."
[57] Lewis, "Humanitarian Theory of Punishment by C.S. Lewis."
[58] Lewis, "Humanitarian Theory of Punishment by C.S. Lewis."

must pursue a scientific objective. Even for the so-called social sciences, scientific objectives must view humans as objects or "cases"—means to an end. If you want less crime, do this. If you want to grow GDP, do that. Want to mitigate a crisis? Here's what needs to be done. There is no such thing as "moral" or "immoral" in science; it recognizes only the efficient and the inefficient, the successful and the ignorant. Plus, if science is the weapon, it needs an ideology to aim it. Ideologies do not see humans; they see only goals, for which science can be employed. For example, the ideology says, "I want a society to be strong," and a (flawed, discredited) science of eugenics says, "then sterilize imbeciles." The natural law, on the other hand, sees people as individual human beings entitled to their own inalienable natural rights.

Natural law cannot aim science because the natural law is not result-oriented. It is human-oriented. It cares only about doing what is "right" in ways that are "just." Under the retributive theory, as C.S. Lewis put it, "to be punished, however severely, because we have deserved it, because we 'ought to have known better,' is to be treated as a human person made in God's image."[59] Of course, science can help the natural law. Forensic science helps provide juries with reliable evidence. Psychology can help explain why a child is less morally culpable than an adult. Medicine can explain how a crime effected a victim. Parole boards might help provide a basis for believing that mercy is justified. But all of these things must still keep intact the humanity of those involved. And the same is true outside of the context of criminal justice. Science can help explain whether an inventor's discovery deserves a patent. Economics can help one understand whether a market is respecting and enhancing liberty. But science is neither the mechanism by which natural law is found, nor the driving force for achieving its ends; the People are.

Up to this point, the focus has been on how a scientifically implemented ideology sees individual people. The answer is that it intrinsically must see them as scientifically manipulatable objects. But scientists are themselves humans. Can't we trust that most scientists are not Nietzschean nihilists or ideologues like Holmes? Even if the People delegate political power to them, might they moderate themselves by operating within the natural law?

Lewis did not think so. In *The Humanitarian Theory of Punishment*, he dismissed the idea quickly: "I will not pause to comment on the simple-minded view of fallen human nature which such a belief implies."[60] He explained this view more thoroughly in his essay *Equality* on why democracy is preferable to rule by philosopher-kings:

---

[59] Lewis, "Humanitarian Theory of Punishment by C.S. Lewis."
[60] Lewis, "Humanitarian Theory of Punishment by C.S. Lewis."

A great deal of democratic enthusiasm descends from the ideas of people like Rousseau, who believed in democracy because they thought mankind so wise and good that everyone deserved a share in the government. The danger of defending democracy on those grounds is that they're not true...I find that they're not true without looking further than myself. I don't deserve a share in governing a hen-roost, much less a nation...The real reason for democracy is just the reverse. Mankind is so fallen that no man can be trusted with unchecked power over his fellows.[61]

And even more importantly, as explored in the following section, science itself would agree with Lewis.

Muting Morality

"Follow the science" is a modern mantra. Science has proven successful. It has exponentially improved our life expectancies and material comforts. "Follow the science" is an expression of faith that science will lead us where we want to go. True, science may allow a malevolent actor to cover his or her true motive. But, by and large, most people, including most scientists, have an innate moral compass. Shouldn't most be able to be trusted to wield science and political power for morally good ends? Perhaps people like C.S. Lewis and America's founders had an overly cynical view of human nature.

In the 1960s, Yale psychologist Stanley Milgram decided to test such a hypothesis scientifically.[62] The famous results of Milgram's experiment have been discussed ever since, even though their interpretation varies. To conduct the experiment, Milgram published an advertisement throughout the New Haven, Connecticut community seeking people to "help us complete a scientific study on memory and learning."[63] When the subjects entered the experiment, they were greeted by a scientist in a lab coat. The scientist explained that his group was investigating whether corporal punishment helps people learn because "almost no truly scientific studies have been made of it in human beings."[64] The subjects were placed into the role of "teachers" and directed to an apparatus that would administer electric shocks to a "learner" (who was an accomplice of the scientist, though this was unknown to the subjects). The experiment within the experiment was a word association memory game. Whenever the learner failed to remember the correct word in the game, the scientist instructed the subject to administer an electrical shock

---

[61] C.S. Lewis, "Equality," *The Spectator* CLXXI, 192 (1943).
[62] S. Milgram, *Obedience to Authority* (New York: Harper & Row, 1974).
[63] Milgram, *Obedience to Authority*.
[64] Milgram, *Obedience to Authority*.

Ideology Fills the Vacuum

to the learner. The learner would behave as if the shocks caused pain so that the subject continued to believe that the shocks were real.

The device operated by the subjects of the experiment had thirty switches by which the shocks were delivered. The switches were labelled with voltage designations ranging between 15 and 450 volts and had "verbal designations" within the range as follows: "Slight Shock, Moderate Shock, Strong Shock, Very Strong Shock, Intense Shock, Extreme Intensity Shock, Danger: Severe Shock, and XXX." The scientist instructed the teacher/subject to administer a shock one level higher each time the learner gave the wrong answer. As the level of shocks increased, the learner would exhibit greater and greater "pain," and even start to beg the teacher/subject not to continue shocking him.

If the teacher/subject turned to the scientist to ask whether to proceed or indicated that he wanted to discontinue, the scientist would respond with one of a few pre-selected "prods." These ranged from "please continue," to "the experiment requires you to continue," to "you have no other choice, you must go on." These dispassionate, scientific-sounding prods contrasted with the scripted protests from the learner. As the shock intensity increased, the learner would say things like, "experimenter, get me out of here! I won't be in the experiment any more! I refuse to go on!," or "I can't stand the pain," or just let out agonized screams.[65]

At 300 volts, the learner would "shout[] in desperation that he would no longer provide answers to the memory test." The scientist would then instruct the teacher/subject to consider a non-response to be a wrong answer and administer the shock anyway. At 315 volts, after a violent scream, the learner would reaffirm that he was no longer a participant. Thereafter, the learner would refuse to say anything and would "shriek[] in agony whenever a shock was administered."[66]

Milgram surveyed audiences on how far they would go in the experiment and how far they thought most others would go. These audiences universally predicted that mercy would be shown to the learner and that the subject would refuse to continue after a low to moderate shock level. On average, psychiatrists predicted that "most subjects would not go beyond the 10th shock level [150 volts], about 4 percent would reach the 20th shock level [300 volts], and about one subject in a thousand would administer the highest shock on the board."

---

[65] Milgram, *Obedience to Authority*
[66] Milgram, *Obedience to Authority*

Thus, the assumption underlying the survey data was that "people are by and large decent and do not readily hurt the innocent."[67]

Milgram's original experiment was published in 1963 using 40 subjects.[68] His results shocked the audiences' expectations. None of Milgram's 40 original subjects refused to continue before administering the 300 volt level of shock. Five refused to obey past the 300-volt level. And another nine broke off before administering the highest shock on the board. Even though most were agitated, disturbed and tense, twenty-six of the forty subjects completed the experiment, administering the highest shock which was described by the apparatus as "XXX."

Some later variations in the experiment changed how the teacher/subject and learner interact. For example, in one iteration, the learner had to touch a shock plate to receive the shock, and at one point of the experiment, the learner refused to continue to touch the plate. The scientist would then instruct the teacher/subject to forcibly hold the learner's hand down on the plate to receive the shock. And, while this lowered the average compliance rate compared to the base study, again a disturbingly large number of subjects continued to forcibly administer the shocks.

The results of Milgram's experiments were no fluke. They have been re-created in many forms over the years in many different, but similar, situations and the results have been largely reproduced. What has differed is the multitude of interpretations given for the results. Milgram himself offered two possible explanations for the result. First, he thought that the results demonstrated people's tendency for conformism, under which they will leave decision making to the hierarchy of a group. Second, he advanced the "agentic state theory," under which people come to view themselves as an instrument for carrying out another person's wishes. As a mere instrument, they no longer feel morally responsible for carrying out another's actions (in this case, those of the scientist instructing them to administer the shocks).

Milgram's interpretations certainly seem to form part of the explanation. Milgram admitted that he designed his experiment to try to determine what made seemingly normal, everyday people in Germany commit heinous acts against their fellow man during World War II. He wondered about examples like that of Adolf Eichmann, whose widely publicized trial in Israel demonstrated to many that he appeared to be just a "normal" person, who would ordinarily be thought incapable of being a prime driving force behind the holocaust.

---

[67] Milgram, *Obedience to Authority*
[68] S. Milgram, "Behavioral Study of Obedience," *Journal of Abnormal and Social Psychology* 67, 371 (1963): p. 371.

Ideology Fills the Vacuum

Hannah Arendt famously described this phenomenon as the "banality of evil," where such results can occur upon a combination of petty motives for career advancement within a hierarchy bent on evil. Milgram's experiment certainly provides some insight into how it is that everyday people can, under certain circumstances, freely commit otherwise unexplainably evil acts. It represented some scientific support for the "I was just following orders" defense heard so frequently at the Nuremberg trials.

But Milgram's explanations do not really account for the "scientific" aspects of his experiment's design. After all, Milgram's flyer soliciting participants notably asked for "help" from the community to complete a "scientific" study. The person directing the experiment was dressed in a lab coat and told the subjects how science had heretofore been unable to complete "truly scientific" experiments on how punishment affects learning. And one of the "prods" used by the scientist during the experiment to get the subject to continue was: "the experiment requires you to continue."

Researchers in 2014 picked up on these aspects of Milgram's original experiment and studied them.[69] By controlling the content of the "prods" administered to the subjects, these researchers sought to determine whether an "order," on the one hand, or an "appeal to science" on the other, would increase the likelihood that the subject would continue in the experiment. On some subjects, they used Milgram's prod that "the experiment requires you to continue," while others were straightforwardly ordered to continue. The researchers found:

> Across all conditions, continuation and completion were positively predicted by the extent to which prods appealed to scientific goals but negatively predicted by the degree to which a prod constituted an order. These results provide no support for the traditional obedience account of Milgram's findings but are consistent with an engaged followership model which argues that participants' willingness to continue with an objectionable task is predicated upon active identification with the scientific project and those leading it.[70]

In other words, the subjects were willing to set aside their moral views, not so much because an authority figure was telling them to conform and continue, but because they identified with the scientific goals of the experiment.

---

[69] S. Haslam, "Nothing by Mere Authority: Evidence that in an Experimental Analogue of the Milgram Paradigm Participants are Motivated not by Orders but by Appeals to Science," *Journal of Social Issues* 70, 473 (2014).

[70] Haslam, "Nothing by Mere Authority: Evidence that in an Experimental Analogue of the Milgram Paradigm Participants are Motivated not by Orders but by Appeals to Science."

In many ways, this explanation for Milgram's results is much more disturbing than those proposed by Milgram. Sure, it is discouraging that people will sometimes "blindly obey" authority figures. But those effects can be controlled so long as authorities and the institutions they run are kept virtuous, in which case blind obedience will be in the service of the good and morally upright. The findings from the 2014 researchers, however, suggest that the subjects yielded to an "engaged fellowship" model of obedience. They obeyed because they identified with the scientific goals of the scientist. They rationalized their actions, and traditional morality took a backseat to the "greater good" of an important scientific enterprise. Psychologist Clifford Scott explained that the Milgram experiments show "the idealism that scientific inquiry had on the volunteers: 'The influence is ideological. It's about what they believe science to be, that science is a positive product: it produces beneficial findings and knowledge to society that are helpful for society. So there's that sense of science providing some kind of system for good.'"[71]

Science should not be elevated from its role as servant in the search for truth to assume a quasi-moral dimension. The directive to "follow the science," if interpreted as a directive for action rather than knowledge, has the potential to mute natural moral conscience. Instead of following the fixed star of moral law, science itself, through the Milgram line of experiments, shows that an "engaged fellowship" will abandon the moral law in favor of what it views as the higher purpose of the scientific enterprise. Cruelty and inhumanity will inevitably ensue.

Fortunately, these scientific findings suggest their own cure for how to preserve one's humanity and avoid a muting of the moral whispers of conscience. While it is anecdotal, one example test run from Milgram's *Obedience to Authority* palpably illustrates what the data made clear. Milgram described the subject in one case as a "gaunt, ascetic man" who taught "Old Testament liturgy" at a divinity school. At first, things in the experiment looked troubling. The subject purposely "adjusted his seat to look at the learner while administering shocks." He seemed "curt and officious" and "appear[ed] almost to be chastising [the learner] for his failing" when the learner missed a word in the memory game. "Each time he administers a shock, his lips are drawn back, and he bares his teeth."[72] It seemed at first that Milgram might have found himself a cruel, religious zealot.

But as the experiment continued, the subject started to question the scientist, becoming "extremely agitated" and wondering whether the scientist had

---

[71] R. Sassower, *Compromising the Ideals of Science* (Palgrave Macmillan, 2015), p. 89.
[72] Milgram, *Obedience to Authority*.

"considered the ethics of this thing." When the experiment reached 150 volts and the learner said that he did not want to continue, the subject balked at going any further. Before the experiment ended, this was how the subject responded to the scientist's prods:

> SCIENTIST: Its absolutely essential to the experiment that we continue.
>
> SUBJECT: I understand that statement, but I don't understand why the experiment is placed above this person's life.
>
> SCIENTIST: There is no permanent tissue damage.
>
> SUBJECT: Well, that's your opinion. If he doesn't want to continue, I'm taking orders from him.
>
> SCIENTIST: YOU have no other choice, sir, you must go on.
>
> SUBJECT: If this were Russia maybe, but not in America. (The experiment is terminated.)[73]

This subject continued to recognize the humanity of the learner, and thereby the learner's inherent authority to choose to discontinue the experiment despite its supposed scientific importance.

After the experiment was discontinued and after the scientist explained the true purpose of the experiment to the subject, the scientist asked him what he thought was the best way to strengthen a person's resistance to inhumane authority. The subject responded, "If one had as one's ultimate authority God, then it trivializes human authority." As noted above, Milgram's thesis was that humans are programmed to follow authority. Accordingly, Milgram personally viewed this example as one where the subject did not "repudiate[e]" authority, but just "substitute[ed]" a good authority for a bad one.[74] But there is more to it than that. The subject in the experiment recognized the authority of the individual person sitting next to him. That authority was based on his humanity and the fact that the moral law precedes and predicates all forms of legitimate human authority and law. The scientist's scientific claim that the machine would cause no permanent tissue damage was seemingly accepted by the subject, but it was rejected as insufficient once the person with authority over the situation (the learner) had made his choice. It is also interesting that the subject affirmed his own authority to make his own choices after the scientist purported to claim that he had none. The subject's repudiation of the scientist's claim that he had "no other choice, sir" was based on the fundamental precepts of America.

---

[73] Milgram, *Obedience to Authority*.
[74] Milgram, *Obedience to Authority*.

Other stories of people who cut off the experiment midway, which were reported by Milgram in his book, share this philosophy. For example, an engineer who emigrated from Holland after World War II called off the experiment at 255 volts, saying "Why don't I have a choice? I came here on my own free will. I thought I could help in a research project. But if I have to hurt somebody to do that...I can't continue. I'm very sorry. I think I've gone too far already, probably."[75] On the other hand, some of those who continued the experiment through to conclusion clearly embodied the more "progressive" and scientific viewpoint. A thirty-nine-year-old social worker did everything the experiment asked of him. He spoke to the learner in a "refined," "authoritative" and "officious[]" manner.[76] He would instruct the learner about the purported consequences of his actions thusly: "Mr. Wallace, your silence has to be considered as a wrong answer." And as the experiment proceeded to higher and higher voltages, the social worker started to nervously laugh, at first behind his hand, but then in an uncontrolled manner. After the experiment, the social worker said that it was "extremely painful" for him to continue administering the shocks. But he explained his actions: "There was I. I'm a nice person, I think...and caught up in what seemed a mad situation and in the interest of science, one goes through with it."[77]

Conclusion

Science has been misused to justify a wide variety of ideologies spanning the political spectrum. It has been employed in the service of socialism and communism as well as eugenics and social Darwinism. It has lobotomized criminals and sterilized "imbeciles." Throughout it, all is a simple underlying premise: that an ideological end can and should be achieved under the efficient eye of a scientist. This premise necessarily disregards the inherent humanity and pre-existing liberty of individuals. Individual people like Carrie Buck, a victim of a warped system who was sacrificed to appease the good social graces of her embarrassed foster parents, will fall through the cracks in such a system.

Today, just as it was in Carrie Buck's case, scientific primacy is ironically touted as more "humanitarian" and "enlightened." This disguises the reality which is precisely the opposite. There is certainly no reason to doubt that such scientific administrators are nice people who have only the best of intentions in mind. As noted at the beginning of the chapter, this discussion has nothing whatsoever to do with everyone's favorite anecdotes about the overtly abusive bureaucrat on a power trip lording over some petty fiefdom. We all know that's

---

[75] Milgram, *Obedience to Authority*.
[76] Milgram, *Obedience to Authority*.
[77] Milgram, *Obedience to Authority*.

wrong. This is about the vast majority of public servants that want nothing more than to help society, to "cure" the drug addict and the sex offender, to protect people from bad choices, and to lend a helping hand to those that have fallen on hard times. To be sure, there is absolutely nothing wrong with those goals. The problem arises when such goals are pursued scientifically and ideologically rather than through the lens of natural law and the natural rights of individuals.

A scientist sees subjects. There are inputs, algorithms, and outputs. A truly scientific approach to a human problem cannot see the humanity of those it seeks to help. It sees an ideological problem to be solved—crime, addiction, poverty—and sets about a "final" solution. Milgram's experiment showed that a scientific endeavor could seemingly justify ordinary people administering purportedly excruciating electrical shocks in the name of improving memory and learning. The ideological end, pursued with scientific efficiency, is able to mute the objections of the individual conscience.

Science is a great servant but a terrible master. The one fixed star that prevents science from breaking its chains is the natural law. A person is naturally endowed with rights to life and liberty. Since liberty was not bestowed by any human authority, be it social, cultural, governmental, and scientific, none can take it away in a manner inconsistent with the natural law. In other words, governmental and scientific authorities must be held morally accountable to the natural law, not to science, to justify their actions. Carrie Buck cannot be sterilized just because the law posits that she can be after a certain proceeding found her to be an "imbecile," incompetent to reproduce. Nor can she be sterilized because science says it will improve our society, our race, or even Carrie Buck's personal happiness. To say these things is to make law and science self-justifying rather than accountable to the higher authority of the natural law and the individual rights of human beings.

The natural law is the one and only *humanitarian* law because it alone asks whether an action is consistent with our humanity and the natural rights, freedoms, and liberties that entails. As the enlightenment philosopher, Immanuel Kant would frame it, the natural law treats the People as ends in themselves. In the eyes of the natural law, someone as seemingly lowly as Carrie Buck is equal in value to someone as venerable, as powerful, as intelligent, and as historically influential and "significant" as an Oliver Wendell Holmes, Jr. By contrast, an ideology treats people as means to an end, even if that ideology is something as seemingly good as science or efficiency or maximizing happiness. It blinds its practitioners to the humanity of those they want to help, and ultimately leads to the loss of the practitioners' humanity through cold, scientific administration.

7.
Conclusion: Science and Liberty

> *Individually and through their professional organizations, scientists and technicians could do a great deal to direct the planning toward humane and reasonable ends. In theory everyone agreed that applied science was made for man and not man for applied science. In practice great masses of human beings have again and again been sacrificed to applied science.*
> —Aldous Huxley, *Science, Liberty and Peace.*

This book is called Science and Liberty because it is not an either/or matter. Nothing here is anti-science. But at the same time, we must not become anti-liberty. I have been arguing for a type of outlook that respects science's proper sphere while still being ordered around the demands of liberty.

Before summarizing the argument in favor of such an outlook, one model for this balanced attitude springs to mind. I became a patent lawyer after receiving an undergraduate degree in chemistry and being named as an inventor on multiple United States patents from my professional work. Now, litigating patent lawsuits for a living, I can think of no better model for my proposal here, of how to accord the proper balance between science and liberty, than America's system of trial by jury.

The Model of the Jury System

So much of our society has changed since America's founding. Science has changed. Culture has changed. The government has changed. Bureaucracies and administrative agencies that the founders could never have imagined were created supposedly to deal with a complex world. Sometimes it can be hard to grasp what the Founders had in mind when they spoke of Liberty and power emanating from "We the People." But, thankfully, what has not really changed over these hundreds of years is America's jury system. There, we can still catch a glimpse of what it means for democratic power to truly be in the service of Liberty.

The right to a jury trial is the only constitutional right that appears explicitly in *both* the original U.S. Constitution and the Bill of Rights.[1] The famous English jurist of the Enlightenment, Sir William Blackstone, called the right to a trial by a jury of one's "peers of every[day] Englishman...the grand bulwark of his liberties."[2] America's Declaration of Independence explicitly complained that King George had deprived the citizens of the colonies of the "Benefits of Trial by Jury." In the Federalist Papers, Alexander Hamilton reported that one of the only things that people on both sides of the Constitutional debate (the federalists and the anti-federalists) could agree upon was the importance of jury trials:

> The friends and adversaries of the plan of the convention, if they agree in nothing else, concur at least in the value they set upon the trial by jury; or if there is any difference between them it consists in this: the former regard it as a valuable safeguard to liberty; the latter represent it as the very palladium of free government...[A]ll are satisfied of the utility of the institution and of its friendly aspect to liberty.[3]

The Founder's reason for placing this extraordinarily high value on juries is precisely the point for our purposes here. The jury is the essence of a democratic structure that serves to secure, protect, and enhance the liberty of the people. The People, embodied directly by the members of the jury, have the political power over the liberty of the individual. Such power simply cannot be abandoned to some "expert." As Justice Antonin Scalia explained in one of his opinions while serving on the Supreme Court:

> [T]he Constitution does not trust judges to make determinations of criminal guilt. Perhaps the Court is so enamoured of judges in general, and federal judges in particular, that it forgets that they (we) are officers of the Government, and hence proper objects of that healthy suspicion of the power of government which possessed the Framers and is embodied in the Constitution. Who knows? 20 years of appointments of federal judges by oppressive administrations might produce judges willing to enforce oppressive criminal laws, and to interpret criminal laws oppressively-at least in the view of the citizens in some vicinages where criminal prosecutions must be brought. And so the people reserved the function of determining criminal guilt to themselves,

---

[1] U.S. Constitution Art. III, § 2, clause 3; Sixth Amendment.
[2] W. Blackstone and R. Kerr, *Commentaries on the laws of England* (London: J. Murray, 1880), vol. 4
[3] A. Hamilton, J. Jay, and J. Madison, *The Federalist Papers* (1788), at Federalist No. 83.

sitting as jurors. It is not within the power of us Justices to cancel that reservation...by reviewing the facts ourselves and pronouncing the defendant without-a-doubt guilty.[4]

I have tried a fair number of cases before a jury. In the courtroom, you can truly feel the raw democratic power of the jurors. Judges profoundly respect them. Lawyers tread carefully in their presence; every strategy, even every word sometimes, is carefully considered and crafted in advance. The parties know that their fates are in the juror's hands. Witnesses know that the jury's keen eye for deception and canny ability to use common sense is always at the ready, and they simply won't be believed if they try any funny business. The same goes for expert, scientific witnesses. Expert witnesses do not talk down to American jurors. Sure, jurors are interested to hear what an expert, especially a scientist, has to say if it is something in their wheelhouse. But our system always allows the jurors to determine for themselves (1) whether what the expert said was truly scientific and within the scope of their expertise, (2) what bearing it has on the case, and (3) what importance it has for the moral or value-based judgments jurors are called upon to make. And then, most importantly, what the jury says about the truth of the matter, goes.

The jury system is the true model for maintaining the balance between Science and Liberty in the broader world. The People wield the power in the manner engineered by the Founders so that it will serve, not undermine, the interests of Liberty. After all, we do not let juries charge and prosecute people or preside over trial proceeding like judges because that could lead to mob justice and tyranny. Similarly, the Founders designed a system for making law where power simultaneously originated in the People, but it was funneled and structured to serve Liberty by avoiding mob rule and majoritarian oppression.

Scientists, experts, and bureaucrats should be valued for their opinions and insights. But no expert witness in any jury trial would ever even think to try to cram those opinions down on a juror. It would rightly be found off-putting and inappropriate. A directive to "follow the science, or else" would quickly be dismissed as out of order and only harm your chances with a jury. Instead, an expert witness must patiently and respectfully explain his or her view to the jury. Reasons must be provided and candidly exposed for fair criticism through cross-examination. If an expert witness thinks something is too complicated to "dumb down" for a jury or too well-accepted among his peers for him to deign to justify it, well...too bad.

Notice also that nothing about America's jury system is anti-science. The expert and the scientist are welcomed into the courtroom. They often become

---

[4] *Neder v. United States*, 527 U.S. 1 (1999) (Scalia, J. dissenting).

a focal point in trials on complex matters. Juries are always interested in what expert witnesses have to say. In the context of criminal law, think of our collective fascination with forensics, DNA matching, and the *CSI*-effect. A diverse array of scientific endeavors have been represented in the courtroom from the hard sciences like chemistry, to the human sciences like psychology, to the social sciences like economics. But that said, no one could ever say that the expert witness who enters the courtroom is in charge. No one says he or she is infallible or that the testimony given must be heeded. A healthy, reasonable skepticism is applied by judge, lawyer, and juror alike. A judge can throw expert testimony out of court by finding it to be un-scientific or by finding the witness lacking in expertise. A lawyer can test the expert and probe for scientific weaknesses and biases through cross-examination. And the jury—the People— have the ultimate say: to believe or not to believe, to weigh the expert's advice on their scales of human values and justice. In the courtroom, none of this is decried as backwards or anti-science. Expert, judge, lawyer, and juror alike are just doing their traditionally assigned job in the service of truth and liberty.

The American and the European Model

The respect and sovereignty accorded to juries in America allows us to glimpse the basic political outlook that real democracy entails. It finds its origins in English law, brought to its fulfillment and culmination in America's constitutional structure beginning with the consent of the governed, "We the People."

But history and the influence of the continental European model has eroded democratic ideals in favor of power being wielded in the hands of the few. They purport to be doing the will of the People, but they also claim that, as experts, they are able to make government more efficient and responsive to the People's will. They do not trust the People to be able to carry out their will within the restrictions of a democratic structure that is designed to avoid majoritarian tyranny.

The Continental model began breaching America's constitutional floodgates in the late Nineteenth and early Twentieth Centuries. A progressive agenda reacted against industrialization in the Gilded Age. But when the laws failed to change as quickly as these reformers believed that they should, they cynically concluded that America's constitutional structure could not adequately respond to what they saw as the clear will of the People. And so, political reform bled into structural reform as Woodrow Wilson and Frank Goodnow borrowed heavily from the Continental model of administrative law. To be sure, they were candid about their disdain for America's founders and the constitutional structure they designed. America's founders were cast as out-of-touch idealists, as greedy capitalists looking to set up a system that would work in their favor, or as quaint relics of a bygone era whose insights were of no relevance to the modern world.

Turning back to the jury model, we can see just how different the viewpoints of America's founders were from the administrators in Continental Europe. The fundamental right to a jury trial is so enmeshed in American culture that we often find it surprising how rare it is in other legal traditions. In France, juries are extremely rare. They are non-existent in civil cases, and only a limited form of jury trial exists for those charged with the most serious of crimes (those punishable by more than 10 years in prison). These serious crimes are tried in the French *cour d'assises* where a panel of six jurors and three judges try the case together. The judges participate in the jurors' deliberations and have equal votes. In other words, these judges are present the whole way through the trial exerting influence and the stature of their position. Germany briefly experimented with juries after the German revolutions in 1848-1849. But "the continental jury never really became acculturated and soon suffered a decline" leading to its abolition.[5] Today, as in France, only serious crimes come with a *quasi*-jury. In Germany, they are called lay judges, but like the French *cour d'assises*, they are always put on panels with traditional, professional judges, who even often outnumber the lay judges. As one scholar summarized, "[t]he prevailing contemporary continental system is that of a unified bench in which the professional judge or judges are flanked by lay assessors. Even in France, after the reforms of 1941, the 'jurors' deliberate and vote with the professional judges, so that the system remains that of 'jury trial' in name only."[6] The liberty of the accused lies in the hands of expert governmental officials and bureaucratic administrators, not the People themselves through their jury representatives.

The continental philosophy also finds a microcosm in its use of an inquisitorial justice system, in contrast to America's adversarial system. In the continental criminal process, a bureaucratic "accuser" (typically a public prosecutor) is tasked by "the investigating judge, or some other purportedly impartial official...with the collection of evidence."[7] The accuser compiles a dossier of evidence, which the defendant is allowed to inspect before the case proceeds to the trial phase. At the trial, "[p]roof-taking is presided over by the professional judge. He is not only very active in questioning witnesses, but is also authorized and required to raise all issues relevant to the charge."[8]

---

[5] M. Damaska, "Evidentiary Barriers to Conviction and Two Models of Criminal Procedure," *University of Pennsylvania Law Review* 121, 506 (1973).
[6] Damaska, "Evidentiary Barriers to Conviction and Two Models of Criminal Procedure."
[7] Damaska, "Evidentiary Barriers to Conviction and Two Models of Criminal Procedure."
[8] Damaska, "Evidentiary Barriers to Conviction and Two Models of Criminal Procedure."

The continental *civil* process involves the parties submitting documents and exhibits to a professional judge. No trials are held; instead, there might be a final "oral hearing" that "might last no longer than twenty to thirty minutes" in which the parties "reiterate their respective requests."[9] Before the final hearing, the judge essentially carries out an investigation or a judge-directed discovery process. "[I]n traditional terms, [this] has been called '*l'instruction*.'"[10] The judge "collects all the materials the parties submit to him, interrogates witnesses, and appoints experts… Experts must be strictly neutral persons in civil-law systems. They are appointed by the judge and paid by the state treasury… It is seldom that the parties insist on questions being posed to the expert at the oral hearing. Not infrequently the parties settle the case after having been informed of the expert's opinion."[11]

At every possible step, the premise of the continental model is that a government official and/or an expert witness is the more efficient and superior mechanism for the administration of justice. This deference is part and parcel of the more general continental culture, especially when a neutral subject matter expert has spoken. As an Italian scholar of comparative law recently put it, "[e]xpert opinions and reports are not formally binding on judges in any European jurisdiction. In the real world, however, expert opinions have a crucial and frequently decisive influence on the final outcome of the process."[12] In Italy, experts are not considered witnesses; they are the judge's auxiliaries. If the judge accepts the expert's opinion, the judge does not have to explain why. Only if the judge dissents from the expert's conclusions does the judge need to give reasons and set forth other grounds for the decision. Of course, this very rarely occurs.

The precise opposite is true in the American model. At every possible step, the premise of the American model is that the People are in charge of finding the truth. Yes, a judge presides to maintain order, to ensure that everyone complies with the procedure, and to bind the parties to the law. But the parties themselves control the allegations, the arguments, and the presentation of the evidence, including the presentation of scientific evidence. And the jury is given plenary, unfettered power to find the facts, to ascertain the truth of the

---

[9] P. Schlosser, "Lectures on Civil Law Litigation Systems and American Cooperation with Those Systems," *U. Kan. L. Rev.* 45, 9 (1996).

[10] Schlosser, "Lectures on Civil Law Litigation Systems and American Cooperation with Those Systems."

[11] Schlosser, "Lectures on Civil Law Litigation Systems and American Cooperation with Those Systems."

[12] P. Monaco, "Scientific Evidence in Civil Courtrooms: A Comparative Perspective," *IUS Gentium* 77, 95 (2020).

Conclusion: Science and Liberty 153

matter. It is truly democratic because the People are trusted and empowered every step of the way.

Conclusion

The preceding chapters established that the continental administrative reforms ushered in by the early Twentieth Century progressives were openly hostile or at least dismissive of the founding philosophy of America. While the incommensurable tension between these two views may be difficult to see today when the borderlines seem to be so blurred, it can be seen clearly in the differences between the American and continental justice systems, especially their approach to juries. But even if you were to accept that the continental model of administrative law was inconsistent with American popular sovereignty, which is truly better? The three preceding chapters of this book gave three reasons to prefer the American philosophy.

First, no matter their trappings in neutrality and science, experts do not have privileged access or insights on values as opposed to raw facts. Neil deGrasse Tyson's dream of a utopian "Rationalia" is like the dream of using concentrated sulfuric acid to wash your clothes. Acid has its uses, and some of its properties might be helpful in a certain proportion to accomplish the task. But taking that insight too far is just going to create a mess. As David Hume showed, an "ought" cannot be derived or inferred from what "is." People take what "is" and overlay it with a map of values and morality. Political power exercised by "experts" is going to necessarily displace the People's values with the values of the expert, or at least what the expert thinks the People's values are (and query whether such an expert will ever think his own values are out of touch with those of the People). Administrative rule is just a preference for philosopher kings. It is not democratic. If a government is supposed to be of the people and for the people, it needs to be by the people.

Second, scientists are fallible, and the People have a legitimate role to play in evaluating and considering scientific claims. With example after example from both the hard and social sciences, it is clear that Sir Francis Bacon's Idols and Richard Feynman's Cargo Cult Science are alive and well today. Scientists are flesh and blood humans like everyone else. They are not immune to the normal pressures of career, to the glitz of making headlines, to the need to fit in and defer to authority, to their own preconceived beliefs, prejudices, and politics, and to advancing their own theories or seeking what they perceive as "the greater good," at the expense of true dispassionate science. Often it is the People and not the scientific establishment that is best positioned to root out these fallacies and failings. Think, for example, of Brian Deer's exposé on the Andrew Wakefield's supposed vaccination study.

Third, and most importantly, human nature requires that political power be left in the hands of the People and not in the hands of the few acting as "experts." As one former President put it at his first Inaugural Address:

> From time to time we've been tempted to believe that society has become too complex to be managed by self-rule, that government by an elite group is superior to government for, by, and of the people. Well, if no one among us is capable of governing himself, then who among us has the capacity to govern someone else?[13]

In saying this, he may have been inspired by one of our greatest presidents' first Inaugural observation: "Why should there not be a patient confidence in the ultimate justice of the people? Is there any better or equal hope in the world?";[14] and by our first President's Farewell warning to beware "that love of power, and proneness to abuse it, which predominates in the human heart."[15]

When science is entwined too closely with political power, it tends to produce ideology. It eschews the natural law that binds us all together as humans and inserts a supposedly scientific lens through which to accomplish some political end. Science, by necessity, can only study and interact with people as objects. It deals in inputs and outputs, objectives and means, causes and effects. As a result, such political objectives are inevitably pursued with cold, inhuman efficiency. When the "science" of eugenics was in vogue, it allowed Carrie Buck to be sterilized. There was no room for discussion of liberty, natural law, or inalienable rights when a political agenda is cast in scientific terms. "Three generations of imbeciles [were] enough" to justify an inhuman action, according to Oliver Wendell Holmes, Jr.[16]

Moreover, the great lesson of the horrors of the Twentieth Century, as confirmed by scientific experiments like those of Stanley Milgram, is that ordinary people can be easily seduced by a scientifically-implemented ideology. It can happen even to the best among us: Oliver Wendell Holmes was no monster and seven of his eight colleagues agreed with his opinion in *Buck v. Bell*. Once the conscience of natural law has been neutralized, almost anything can be

---

[13] Avalon Project, "First Inaugural Address of Ronald Reagan," Avalon.law.yale.edu, Accessed: May 28, 2021, https://avalon.law.yale.edu/20th_century/reagan1.asp
[14] Avalon Project, "First Inaugural Address of Abraham Lincoln, March 4, 1861," Avalon.law.yale.edu, Accessed: May 28, 2021, https://avalon.law.yale.edu/19th_century/lincoln1.asp
[15] Our Documents, "Transcript of President George Washington's Farewell Address (1796)," Ourdocuments.gov, Accessed: May 28, 2021, https://www.ourdocuments.gov/doc.php?flash=false&doc=15&page=transcript
[16] *Buck v. Bell*, 274 U.S. 200 (1927).

rationalized in the service of a scientific end. Consider the case of Maximilien Robespierre and his Reign of Terror. He had a vision for a "free" society that was based on rationality (recall the Temple of Reason and the French Republican calendar) and that provided rights and material necessities to its people. His Committee of Public Safety was given plenary political and police powers to guide the fledgling French republic rationally and scientifically. In defense of this, he said, "everywhere we must level the obstacles and hindrances to the execution of the wisest measures" of government.[17] But as it always goes when the natural law is ignored, the artificially constructed dream turns to a nightmare, and the very instrument of execution abused so easily by Robespierre was just as easily turned upon him after a feigned trial.

The blessing and the curse of democracy is that the People have the power. The *hoi polloi* can be just as tyrannical as a king or a Robespierre. It unjustly killed Plato's teacher, which may have been a reason why he thought only the few could rule as philosopher kings. But America's founders discovered a way to tame these unjust propensities of the People by funneling power through a constitutional structure that would keep the focus of law consistent with natural law in securing the People's Liberty. Plato's student, Aristotle, speculated that there might be a collective wisdom of the People that, if properly channeled, would exceed the knowledge, wisdom, and capabilities of any individual "expert":

> The principle that the multitude ought to be supreme rather than the few best ...seems to contain an element of truth....[I]f the people are not utterly degraded, although individually they may be worse judges than those who have special knowledge—as a body they are as good or better.[18]

America proved that Aristotle was right. A nation founded on, and ordered around, democratic Liberty became the greatest the world had ever known. But far from being the inevitable culmination of those civilizations which came before, history has shown that the result in America was inconceivably unlikely. It required a nugget of English Enlightenment, arising from an unusually weak monarchy, to be transplanted and watered in a New World to grow into a fragile republic protected by its remote locale. While attempted in France, it could not be recreated. Other philosophies quickly confused and choked the French revolution. The promise of liberty through self-rule has always been tenuous.

The factional interests of the few or even of a majority will be stifled and softened by the Constitution. Other modes of administration will promise

---

[17] M. Robespierre, *For the Defense of the Committee of Public Safety* (1793).
[18] Aristotle and B. Jowett, trans., *Aristotle's politics and poetics* (New York: Viking Press, 1974), Politics III. 11.

factions that their interests can be achieved faster, more efficiently, and more wisely than can be done through the Constitution. It is, of course, tempting and easy to justify this under the modern banner of science. But it is just the same old Platonic promise of the philosopher king in new garb. The harm to liberty and the humanity of individuals will always be the same. Tempting as it is, such delegations of power under the cover of "science" must be resisted. As one Supreme Court Justice memorably put it: "Frequently an issue of this sort will come before [us] clad, so to speak, in sheep's clothing: the potential of the asserted principle to effect important change in the equilibrium of power is not immediately evident, and must be discerned by a careful and perceptive analysis. But this wolf comes as a wolf."[19]

---

[19] *Morrison v. Olson*, 487 U.S. 654 (1988) (Scalia, J. dissenting).

Bibliography

A.L.A. Schechter Poultry Corporation v. United States, 295 U.S. 495 (1935).

Abrams v. United States, 250 U.S. 616 (1919).

Adorno, T. and M. Horkheimer. *Dialectic of Enlightenment.* New York: Herder and Herder, 1947.

Allen, Christopher P. G., and David M. A. Mehler. "Open Science Challenges, Benefits and Tips in Early Career and Beyond." PsyArXiv (17 Oct. 2018).

Alschuler, A. "The Changing Purposes of Criminal Punishment: A Retrospective on the Law Century and Some Thoughts about the Next." *University of Chicago Law Review* 70, 1 (2003).

Anderson, V. "The present status of the oil shale industry." *Colorado School of Mines Quarterly* 16, 12 (1921).

Aristotle and B. Jowett, trans. *Aristotle's politics and poetics.* New York: Viking Press, 1974.

Artherton, K. "Neil deGrasse Tyson Doubles Down On Rationalia." Popular Science, Aug. 8, 2016. (Accessed: May 28, 2021). https://www.popsci.com/neil-degrasse-tyson-doubles-down-on-rationalia/

Artherton, K. "Neil deGrasse Tyson's Proposed "Rationalia" Government Won't Work." Popular Science, June 29, 2016. (Accessed: May 28, 2021). https://www.popsci.com/neil-degrasse-tyson-just-proposed-government-that-doesnt-work/

Avalon Project. "English Bill of Rights of 1689." Avalon.law.yale.edu. (Accessed: May 28, 2021). https://avalon.law.yale.edu/17th_century/england.asp

Avalon Project. "First Inaugural Address of Abraham Lincoln, March 4, 1861." Avalon.law.yale.edu. (Accessed: May 28, 2021). https://avalon.law.yale.edu/19th_century/lincoln1.asp

Avalon Project. "First Inaugural Address of Ronald Reagan." Avalon.law.yale.edu. (Accessed: May 28, 2021) https://avalon.law.yale.edu/20th_century/reagan1.asp

Bäcklin, E. "Eddington's Hypothesis and the Electronic Charge." *Nature* 123 (1929), pp. 409–410.

Bacon, F. *The Novum Organum of Sir Francis Bacon, Baron of Verulam, Viscount St. Albans, Epitomiz'd for a Clearer Understanding of his Natural History.* 1620.

Bardi, U. "Peak Oil, 20 years later: Failed prediction or useful insight." *Energy Research & Social Science* 48, 257 (2019).

Bastasch, M. "30 Years Ago Officials Predicted The Maldives Would Be Swallowed By The Sea. It Didn't Happen." The Global Warming Policy Forum, 2018. (Accessed: May 28, 2021). https://www.thegwpf.com/30-years-ago-officials-predicted-the-maldives-would-be-swallowed-by-the-sea-it-didnt-happen/

Beard, C. A. *An Economic Interpretation of the Constitution of the United States,* New York: MacMillan, 1913.

Benatar, D. *Better Never to Have Been: the Harm of Coming Into Existence.* Oxford: Oxford University Press, 2006.

Bentham, J. *Introduction to the Principles of Morals and Legislation.* Dover Publications, 2012.

"Biden says he would shut the U.S. down if recommended by scientists." Cbsnews.com. (Accessed: May 28, 2021). https://www.cbsnews.com/news/biden-shut-us-down-coronavirus-if-recommended-scientists/

Birge, R. "Probable Values of the General Physical Constants." *Review of Modern Physics* 33 (1929): pp. 233-39.

Blackstone, W. and Kerr, R. *Commentaries on the laws of England.* London: J. Murray, 1880.

Bourne, E. "Alexander Hamilton and Adam Smith." *The Quarterly Journal of Economics* 8, 3 (1894): p.328.

Buck v. Bell, 274 U.S. 200 (1927).

Bullock, C. *Essays on the monetary history of the United States.* FORGOTTEN Books, 2015.

Burke, E. *Reflections on the Revolution in France.* 1790.

Cohen, J. "The Earth is Round (p<.05)." *American Psychologist* 49, 12 (1994): pp. 997-1003.

Coke, E. *The First Part of the Institutes of the Laws of England.* 1628.

Collins, J. *A History of Modern European Philosophy.* Bruce Publishing, 1965.

Columbia University. "Declaration Rights of Man 1793." Columbia.edu. (Accessed: May 28, 2021). http://www.columbia.edu/~iw6/docs/dec1793.html

Croteau, D. "An Analysis of the Arguments for the Dating of the Fourth Gospel." Liberty University Faculty Publications Paper 118 (2003).

Cushing, H. *The Writings of Samuel Adams.* New York: G.P. Putnam's Sons, 1904.

Damaska, M. "Evidentiary Barriers to Conviction and Two Models of Criminal Procedure." *University of Pennsylvania Law Review* 121, 506 (1973).

De Tocqueville, A. *Democracy in America.* London: Penguin Books, 2003.

Delafosse, H. *Le Quatrieme Evangile.* Paris, 1925.

Dewey, J. *Liberalism and social action.* New York: Putnam, 1935.

Don, K. "The Moral Landscape: Why science should shape morality." Salon, Oct. 17, 2010 (Accessed: May 28, 2021) https://www.salon.com/2010/10/17/sam_harris_interview/

Dred Scott v. Sandford, 60 U.S. (19 How.) 393 (1857).

Ehrlich, P. *The Population Bomb: Population Control or Race to Oblivion?.* The Sierra Club, 1968.

Ehrman, B. *How Jesus Became God.* New York: HarperOne, 2015.

Ehrman, B. *Misquoting Jesus: The Story Behind Who Changed the Bible and Why.* New York: HarperOne, 2007.

Einstein, A. "Physics and reality." *Journal of the Franklin Institute* 221, 3 (1936): pp.349-382.

Einstein, A., B. Podolsky, and N. Rosen. "Can Quantum-Mechanical Description of Physical Reality Be Considered Complete?." *Physical Review* 47, 777 (1935).

Eliot, T.S. *The Waste Land.* 1922.

Emerson, R.W. *The American Scholar.* 1837.

Farrington, B. and F. Bacon. *The Philosophy of Francis Bacon.* Chicago: University of Chicago Press, 1964.

Federal Trade Commission v. Ruberoid Co., 343 U.S. 470, 487-88 (1952).

Feynman, R. and E. Hutchings. *Surely you're joking, Mr. Feynman!.* New York: W. W. Norton, 1997.

Founders Online. "From George Washington to Alexander Hamilton, 10 July 1787." Founders.archives.gov. (Accessed: May 28, 2021). https://founders.archives.gov/documents/Washington/04-05-02-0236

Founders Online. "From George Washington to Lafayette, 7 February 1788." Founders.archives.gov. (Accessed: May 28, 2021). https://founders.archives.gov/documents/Washington/04-06-02-0079

Founders Online. "From Thomas Jefferson to Diodati, 3 August 1789." Founders.archives.gov. (Accessed: May 28, 2021). https://founders.archives.gov/documents/Jefferson/01-15-02-0317

Founders Online. "From Thomas Jefferson to John Adams, 4 September 1823." Founders.archives.gov. (Accessed: May 28, 2021). https://founders.archives.gov/documents/Jefferson/98-01-02-3737

Founders Online. "Thomas Jefferson to John Adams, 5 July 1814." Founders.archives.gov. (Accessed: May 28, 2021). https://founders.archives.gov/documents/Jefferson/03-07-02-0341

Franklin, B. *Comparing the Ancient Jews to the Antifederalists.* 1788.

Galarraga, V. and P. Boffetta. "Coffee Drinking and Risk of Lung Cancer—A Meta-Analysis." *American Association for Cancer Research* 25, 6 (2016): pp. 951-957.

George, H. *Progress and Poverty: An Inquiry into the Cause of Industrial Depressions and of Increase of Want with Increase of Wealth: The Remedy.* London: W. Reeves, 1884.

Gompers v. United States, 233 U.S. 604 (1914).

Goodnow, F. *Politics and Administration: A Study in Government.* New York: MacMillan Co., 1900.

Goodnow, F. *Social reform and the constitution.* New York: Franklin, 1911.

Goodnow, F. *The American conception of liberty and government.* Providence: Standard Printing Company, 1916.

Gould, S.J. "Nonoverlapping Magisteria." *Natural History* 106 (March 1997): pp. 16-22.

Gour, H. *The Spirit of Buddhism.* New Delhi: Cosmo, 1990.

GW Regulatory Studies Center. "Reg Stats." Regulatorystudies.columbian.gwu.edu. (Accessed: May, 28, 2021) https://regulatorystudies.columbian.gwu.edu/reg-stats

Hahn, S. and B. Wiker. *Politicizing the Bible: The Roots of Historical Criticism and the Secularization of Scripture.* Germany: Herder & Herder, 2013.

Hamilton, A., J. Jay, and J. Madison. *The Federalist Papers.* 1788.

Harris, S. *The Moral Landscape: How Science Can Determine Human Values.* New York: Free Press, 2010.

Haslam, S. "Nothing by Mere Authority: Evidence that in an Experimental Analogue of the Milgram Paradigm Participants are Motivated not by Orders but by Appeals to Science." *Journal of Social Issues* 70, 473 (2014).

Hays, R. *Echoes of Scripture in the Gospels.* Baylor University Press, 2016.

Holmes, Jr. O.W. "Ideals and Doubts." *Illinois Law Review* 10, 1 (1915).
Holmes, Jr. O.W. "The Path of the Law." *Harvard Law Review* 10, 457 (1897).
Holmes, Jr. O.W. Letter to Emily Hallowell, November 16, 1862.
Holmes, Jr. O.W. Letter to Harold Laski, April 29, 1927.
Holmes, Jr. O.W. Letter to Harold Laski, December 17, 1925.
Holmes, Jr. O.W. Letter to Harold Laski, July 23, 1925.
Holmes, Jr. O.W. Letter to Lady Leslie Scott, May 17, 1912.
Holmes, Jr. O.W. Letter to Lewis Einstein, May 19, 1927.
Holmes, Jr. O.W. *The Common Law.* 1881.
Holmes, Jr. O.W. *The Soldier's Faith.* 1895.
Hume, D. *A treatise of human nature.* Edited by D. Norton and M. Norton. Oxford: Clarendon Press, 2014.
Humphrey's Executor v. United States, 295 U.S. 602 (1935).
Huxley, A. *Science, Liberty and Peace.* Harper & Brothers, 1946.
Ioannidis, J.P.A. "Why Most Published Research Findings Are False." PLoS Med 2, 8 (2005).
Jensen, M. *The new nation.* New York: Vintage Books, 1950.
Kant, I. *Metaphysical Elements of Justice.* Hackett Publishing, 1990.
Kesey, K. *One flew over the cuckoo's nest.* New York: Signet, 1963.
Kirkland, E. *Industry comes of age: Business, Labor, and Public Policy.* New York: Holt, Rinehart and Winston, 1961.
Kojevnikov, A. "The Phenomenon of Soviet Science." *Osiris* 23, 1 (2008): pp.115-135.
Labedz, L. and A. Solzhenitsyn. *Solzhenitsyn.* New York: Harper & Row, 1971.
Lamb, D. *Text, Context and the Johannine Community: A Sociolinguistic Analysis of the Johannine Writings.* A&C Black, 2014.
Lewis, C.S. "Equality." *The Spectator* CLXXI, 192 (1943).
Lewis, C.S. "Humanitarian Theory of Punishment by C.S. Lewis." Matiane, 1953. (Accessed: May 28, 2021) https://matiane.wordpress.com/2018/10/24/humanitarian-theory-of-punishment-by-c-s-lewis/
Locke, J. *Two treatises of government.* Dublin: William M'Kenzie, No. 33, College-Green, 1794.
Loisy, A. *The Origins of the New Testament.* France, 1936.
Lomborg, B. "Al Gore's Climate Sequel Misses a Few Inconvenient Facts." *Wall Street Journal*, July, 27, 2017. (Accessed: May 28, 2021) https://www.wsj.com/articles/al-gores-climate-sequel-misses-a-few-inconvenient-facts-1501193349
Lynch, M. "What Ever Happened To Peak Oil?." *Forbes*, June 29, 2018. (Accessed: May 28, 2021) https://www.forbes.com/sites/michaellynch/2018/06/29/what-ever-happened-to-peak-oil/?sh=12406fdf731a
Lyotard, J. and G. Bennington. *The Postmodern Condition: A Report on Knowledge.* Minneapolis: Univ. of Minnesota Press, 2010.
Mance, H. "Britain has had enough of experts, says Gove." Ft.com, June 3, 2016 (Accessed: May 28, 2021) https://www.ft.com/content/3be49734-29cb-11e6-83e4-abc22d5d108c.

Massachusetts Historical Society. "Adams Papers Digital Edition." Masshist.org. (Accessed: May, 28, 2021) http://www.masshist.org/publications/adams-papers/index.php/view/PJA04dg2

Meeks, W. and J. Bassler. *The HarperCollins Study Bible*. San Francisco: Harper, 1997.

Milgram, S. "Behavioral Study of Obedience." *Journal of Abnormal and Social Psychology* 67, 371 (1963).

Milgram, S. *Obedience to Authority*. New York: Harper & Row, 1974.

Milton, J. *Paradise Lost*. 1667.

Mistretta v. United States, 488 U.S. 361 (1989).

Moffatt, J. "Ninety Years After: A Survey of Bretschneider's "Probabilia" in Light of Subsequent Johannine Criticism." *The American Journal of Theology* 17, 3 (1913): p. 375.

Monaco, P. "Scientific Evidence in Civil Courtrooms: A Comparative Perspective." *IUS Gentium* 77, 95 (2020).

Montesquieu, C. *The spirit of laws*. Glasgow: J. Duncan & Son, J. & M. Robertson, and J. & W. Shaw, 1793.

Morrison v. Olson, 487 U.S. 654 (1988).

Myers v. United States, 272 U.S. 52 (1926).

Neder v. United States, 527 U.S. 1 (1999)

New York Times. "The End of Civilization Feared by Biochemist." Nytimes.com, 1970. (Accessed: May 28, 2021) https://www.nytimes.com/1970/11/19/archives/the-end-of-civilization-feared-by-biochemist.html

Olmstead v. United States, 277 U.S. 438 (1928).

Oreskes, N. "Scientists Get Things Wrong. But We Should Still Trust Science." Time.com, 2019. (Accessed: May 28, 2021) https://time.com/5709691/why-trust-science/

Oreskes, N. *Why Trust Science?*. Princeton University Press, 2019.

Our Documents. "Transcript of President George Washington's Farewell Address (1796)." Ourdocuments.gov. (Accessed: May 28, 2021). https://www.ourdocuments.gov/doc.php?flash=false&doc=15&page=transcript

Pashler & Wagenmakers. "Editors' Introduction to the Special Section on Replicability in Psychological Science: A Crisis of Confidence?" *Perspectives on Psychological Science* 7, 6 (2012).

Pasotti, J. "Maldives Experience That Sinking Feeling." Science, 2005. (Accessed: May, 28, 2021). https://www.sciencemag.org/news/2005/06/maldives-experience-sinking-feeling

"Patrick Henry Speech Before Virginia Ratifying Convention - Teaching American History." Teachingamericanhistory.org. (Accessed May 28, 2021) teachingamericanhistory.org/library/document/patrick-henry-virginia-ratifying-convention-va/

Pestritto, R. "The Birth of the Administrative State: Where It Came From and What It Means for Limited Government." *First Principles Series* 16 (2007).

Petley, B. *The Fundamental Physical Constants and the Frontiers of Metrology*. Adam Hilger, 1985.

Pew Research Center Science & Society. "Trust and Mistrust in Americans' Views of Scientific Experts." Pewresearch.org. (Accessed: May 28, 2021) https://www.pewresearch.org/science/2019/08/02/trust-and-mistrust-in-americans-views-of-scientific-experts/

Pinker, S. *Enlightenment now*. London: Penguin Books, 2019.

Planned Parenthood of Southeaster Pa. v. Casey, 505 U.S. 833 (1992).

Plato. *Plato: The Republic*. Edited by B. Jowett. Charleston: Forgotten Books, 2008.

Pound, R. "Book Review." *American Political Science Review* 3, 281 (1909).

Prescod-Weinstein, C. "Making Black Women Scientists under White Empiricism: The Racialization of Epistemology in Physics." *Signs: Journal of Women in Culture and Society* 45, 2 (2020): pp.421-447.

Raiders of the Lost Ark. Directed by S. Spielberg. Paramount Pictures, 1981.

Randall, T. and H. Warren. "Peak Oil Era Is Suddenly Upon Us." Bloomberg.com, Dec. 1, 2020. (Accessed May 28, 2021) https://www.bloomberg.com/graphics/2020-peak-oil-era-is-suddenly-upon-us/

Rao, T. S. Sathyanarayana and Chittaranjan Andrade. "The MMR Vaccine and Autism: Sensation, Refutation, Retraction, and Fraud." *Indian J. Psychiatry* 53, 2 (2011): pp. 95-96.

Ratzinger, J. *Jesus of Nazareth: From the Baptism in the Jordan to the Transfiguration*. Doubleday, 2007.

"Read the full transcript of Joe Biden's ABC News town hall." ABCnews.com, October 15, 2020. (Accessed: May 28, 2021). https://abcnews.go.com/Politics/read-full-transcript-joe-bidens-abc-news-town/story?id=73643517.

Ridley, M. "Apocalypse Not: Here's Why You Shouldn't Worry About End Times." Wired, 2012. (Accessed: May 28, 2021) https://www.wired.com/2012/08/ff-apocalypsenot/

Roberts, C.H. *An Unpublished Fragment of the Fourth Gospel in the John Rylands Library*. Manchester University Press, 1935.

Robespierre, M. *For the Defense of the Committee of Public Safety*. 1793.

Robespierre, M. *On Subsistence Goods*. 1792.

Robespierre, M. *On the Principles of Political Morality*. 1794.

Robespierre, M. *Speech at 1793 Festival of the Supreme Being*. 1793.

Rohrer, H., 2012. "The Misconduct of Science?." *Proceedings of Trust, Confidence, and Scientific Research*, 2021. (Accessed: May 28, 2021) http://www.abc.net.au/science/articles/2012/07/16/3546732.htm?site=science_dev&

Rosen, J., "Brandeis's Seat, Kagan's Responsibility" New York Times, July 3, 2010. (Accessed: May 28, 2021) https://www.nytimes.com/2010/07/04/opinion/04rosen.html

Rousseau, J. J. *Discourse On The Origin Of Inequality*. 1754.

Rugg, A. "William Cushing." *The Yale Law Journal* 30, 2 (1920): p.128.

Salart, D. "Testing Spooky Action at a Distance." *Nature* 454, 861 (2008).

San Bernardino Sun. "Senate Committee Given Report on Threatened Petroleum Shortage." San Bernardino Sun, Mar. 10, 1937.

Sassower, R. *Compromising the Ideals of Science*. Palgrave Macmillan, 2015.

Scalia, A. "Opening Statement on American Exceptionalism to the Senate Judiciary Committee." Govinfo.gov, 2011. (Accessed: May 28, 2021) https://www.govinfo.gov/content/pkg/CDOC-114sdoc12/pdf/CDOC-114sdoc12.pdf

Schenck v. United States, 249 U.S. 47 (1919).

Schlosser, P. "Lectures on Civil Law Litigation Systems and American Cooperation with Those Systems." *U. Kan. L. Rev.* 45, 9 (1996).

Schopenhauer, A. *The world as will and representation.* 1818.

Schweitzer, A. *The Quest for the Historical Jesus.* Translated by W. Montgomery. Adam and Charles Black, 1910.

Scott, E.H. *Alexander Hamilton, John Jay, James Madison and Other Men of Their Time, The Federalist and Other Contemporary Papers on the Constitution of the United States.* New York: Scott, Foresman and Company, 1894.

Scott, M. *The Essays of Francis Bacon.* New York: Charles Scribner's Sons, 1908.

Senn v. Tile Layers Protective Union, 301 U.S. 468 (1937).

Sheldrake, R. *Science set free.* New York: Deepak Chopra Books, 2013.

Simpsons, Treehouse of Horror VI. [video] Directed by B. Anderson and D. Mirkin, 1995.

Singh, S. *Big Bang.* Milan: Mondolibri, 2005.

Smith, K. "Profits of doom." New York Post, May 29, 2011. (Accessed May 28, 2021) https://nypost.com/2011/05/29/profits-of-doom/

"Speech on the 150th Anniversary of the Declaration of Independence" Teachingamericanhistory.org. (Accessed May 28, 2021) teachingamericanhistory.org/library/document/speech-on-the-occasion-of-the-one-hundred-and-fiftieth-anniversary-of-the-declaration-of-independence/

Southern Pacific Co. v. Jensen, 244 U.S. 205 (1917).

Stevens, A. "Governments cannot just 'follow the science' on COVID-19." *Nature Human Behaviour* 4, 6 (2020): pp.560-560.

Subramaniam, B. "Snow Brown and the Seven Detergents: A Metanarrative on Science and the Scientific Method." *Women's Studies Quarterly* 28, 1/2 (2000).

Taft, W.H., Letter to W. M. Bullitt, Nov. 4, 1926.

Thompson, P. "Silent Protest: A Catholic Justice Dissents in Buck v. Bell." *The Catholic Lawyer* 43, 125 (2004).

Thucydides. "Pericles Funeral Oration for Athenian War Dead." Rjgeib.com. (Accessed May 28, 2021) https://www.rjgeib.com/thoughts/athens/athens.html

Tretkoff, E. "This Month in Physics History: August 1913: Robert Millikan Reports His Oil Drop Results." *American Physical Society* 15, 8 (2006).

Turner, T. *The Vindication of a Public Scholar.* Earth Island Journal, 2009. (Accessed May 28, 2021) https://www.earthisland.org/journal/index.php/magazine/entry/the_vindication_of_a_public_scholar/

Twain, M. "Chapters from My Autobiography." *North American Review*, 1906. (Accessed May 28, 2021) http://www.gutenberg.org/files/19987/19987.txt

United States Census Office. "1890 Census Bulletin (11th Census), Statistics of Manufactures." October 22, 1892.

United States Constitution.

United States Declaration of Independence.

United States v. Motlow, 10 F.2d 657 (7th Cir. 1926).

Urofksy, M. *Biographical Encyclopedia of the Supreme Court: The Lives and Legal Philosophies of the Justices*. CQ Press, 2006.

Vannatta, S. and A. Mendenhall. "The American Nietzsche? Fate and Power in the Pragmatism of Justice Holmes." *UMKC Law Rev.* 85, 187 (2017).

Vinas, M. "Mass gains of Antarctic Ice Sheet greater than losses, NASA study reports." Phys.org., 2015. (Accessed May 28, 2021) https://phys.org/news/2015-10-mass-gains-antarctic-ice-sheet.html

Wakefield, A. et al. "Ileal-lymphoid-nodular hyperplasia, non-specific colitis, and pervasive developmental disorder in children." *The Lancet* 351, 9103 (1998): pp. 637-41.

Waldo, D. *The Administrative State*. The Ronald Press Co., 1948.

Wallace, D. "John 5,2 and the Date of the Fourth Gospel." *Biblica* 71, 2 (1990): p. 177.

Waters, H. "Why Didn't the First Earth Day's Predictions Come True? It's Complicated." Smithsonian Magazine, 2016. (Accessed: May 28, 2021) https://www.smithsonianmag.com/science-nature/why-didnt-first-earth-days-predictions-come-true-its-complicated-180958820/

Watson, J. "Psychology as the Behaviorist Views It." *Psychological Review* 20, 158 (1913).

Webb, R. and J. Rousseau. *Jean Jacques Rousseau: The Father of Romanticism*. F. Watts, 1970.

Werleman, C.J. "How America's Biblical ignorance enables the Christian right." Salon, 2014. (Accessed: May 28, 2021) https://www.salon.com/2014/07/10/how_americas_ignorance_about_the_bible_allowed_the_christian_right_to_push_its_extreme_agenda_partner/

White, D. "The unmined supply of petroleum in the United States." *Transactions of the Society of Automotive Engineers* 14, 227 (1919).

Whitehead, A., D. Griffin, and D. Sherburne. *Process and reality*. New York: Free Press, 1979.

Wilber, K. *Quantum Questions*. Shambala, 1984.

Wills, G. *Explaining America*. New York: Penguin Books, 2001.

Wilson, W. "The Study of Administration." *Political Science Quarterly* 2, 2 (1887): p.197.

Wilson, W. *Congressional Government*. 1885.

Yarbrough, R. "Should Evangelicals Embrace Historical Criticism? The Hays-Ansberry Proposal." *Themelios* 39, 1 (2014): pp. 37-52.

Yates, V. "The Racism in Science's DNA." Medium, 2019. (Accessed May 28, 2021) https://eidolon.pub/the-racism-in-sciences-dna-e82bb7638c35

Yin, J. et al. "Bounding the Speed of Spooky Action at a Distance." *Phys. Rev. Lett.* 110, 260407 (2013) arXiv: 1303.0614.

Zuberi, T. and E. Bonilla-Silva. *White Logic, White Methods: Racism and Methodology*. Lanham: Rowman & Littlefield, 2008.

Index

A

Adams, John, 15, 39, 40, 159
Adams, Samuel, 14, 15, 158
Adorno, Theodor, 79, 157
Age of Enlightenment, 3, 18, 33, 67, 132
An Inconvenient Truth, 118
Ancien Régime, 35
Aristotle, 2, 9, 17, 19, 23, 25, 59, 75, 77, 80, 82, 155, 157

B

Bacon, Sir Francis, 10, 75, 76, 78, 81, 87, 88, 89, 90, 94, 97, 98, 106, 109, 111, 112, 115, 121, 153, 157, 159, 163
Benatar, David, 71
Better Never to Have Been: the Harm of Coming Into Existence, 71
Bible, Historical Criticism, 102
Blackstone, Sir William, 26, 125, 132, 148, 158
Buck v. Bell, 128
Burke, Edmund, 40, 158
Butler, Pierce, 130, 131

C

Cargo Cult Science, 91, 100, 110, 153
Carrie Buck, 127
Checks-and-Balances, 30

Civil War, The, 8, 18, 28, 41, 42, 43, 123, 127
Coke, Sir Edward, 18, 125, 158
Constitutional Convention, The, 28
Coolidge, Calvin, 53

D

De Tocqueville, Alexis, 13, 45, 46, 158
Declaration of the Rights of Man and of the Citizen of 1793, 36
Deer, Brian, 94
Dewey, John, 59
Dialectic of Enlightenment, 79
Dred Scott, 42

E

Ehrman, Bart, 102, 103, 158
Einstein, Albert, 63, 74, 75, 76, 77, 78, 81, 82, 86, 101, 107, 112, 113, 114, 115, 129, 158, 160
Eliot, T.S., The Wasteland, 33
Emerson, Ralph Waldo, 42, 43, 123, 158
Engaged Fellowship, 142
English Bill of Rights of 1689, 18, 19
Enlightenment NOW: The Case for Reason, Science, Humanism, and Progress, 67
EPR Paradox, 113

F

Federalist No. 10, 22, 23, 24, 25, 26
Federalist No. 51, 20, 22
Federalist Number 10, 60
Federalist Papers, The, 6, 15, 20, 21, 22, 23, 24, 25, 26, 27, 29, 34, 38, 39, 54, 148, 159
Feynman, Richard, 91, 92, 97, 100, 110, 111, 153, 159
Frankfurt School, The, 79, 80, 82
Franklin, Benjamin, 29
French Republican Calendar, The, 36

G

General Will, The, 35
George, Henry, 44
Gilded Age, The, 43, 44, 150
Global Warming, 117
Goodnow, Frank, 49, 50, 51, 52, 56, 150, 159

H

Hamilton, Alexander, 6, 15, 19, 20, 21, 22, 23, 24, 25, 26, 27, 28, 29, 34, 39, 54, 148, 158, 159, 163
Harris, Sam, 65, 66, 67, 68, 69, 70, 71, 72, 73, 74, 75, 76, 77, 78, 81, 132, 159
Henry, Patrick, 13, 14, 15, 19, 25, 161
hoi polloi, 1, 2, 127, 155
Holmes Jr., Oliver Wendell, 123, 124, 125, 126, 127, 128, 129, 130, 131, 133, 134, 137, 145, 154, 160, 164
Horkheimer, Max, 79, 80, 81, 157
Hume, David, 10, 18, 34, 68, 69, 70, 71, 153, 160

Humphrey's Executor v. United States, 55
Huxley, Aldous, 147

I

Idol of the Cave, 10
Idol of the Marketplace, 10, 109, 111
Idol of the Theater, 10, 114, 115, 118, 121
Idol of the Tribe, 10, 96
Idols of the Cave, The, 98
Idols of the Marketplace, The, 106
Idols of the Theater, The, 111
Idols of the Tribe, The, 90
Industrial Revolution, 3

J

Jacobins, The, 37
Jefferson, Thomas, 16, 19, 39, 40, 51, 159

K

Kant, Immanuel, 73, 120, 132, 145, 160

L

Lewis, C.S., 129, 135, 136, 137, 138, 160
Lincoln, Abraham, 31, 154, 157
Locke, John, 18, 19, 160
Lyotard, Jean-François, 77

M

Madison, James, 6, 15, 20, 21, 22, 23, 24, 25, 26, 27, 29, 36, 39, 54, 59, 60, 61, 148, 159, 163

Index

Magna Carta, The, 18
Malthusian Theory, 116
Milgram, Stanley, 11, 138, 139, 140, 141, 142, 143, 144, 145, 154, 159, 161
Millikan, Robert Andrews, 99, 100, 101, 163
Milton, John, Paradise Lost, 85
Money Printing, 25
Montagnards, The, 36, 37
Montesquieu, Charles-Louis de Secondat, 19, 25, 161
Myers v. United States, 53

N

Novum Organum, 75, 88

O

One Flew over the Cuckoo's Nest, 123
Oreskes, Naomi, 92
Origins of American Progressivism, 43

P

Peak Oil, 118
Pericles, 2, 17, 163
philosopher-kings, 10, 16, 18, 20, 27, 37, 39, 137
Pinker, Steven, 3, 67, 68, 74, 75, 77, 78, 81, 162
Plato, 2, 3, 6, 9, 10, 16, 17, 19, 20, 27, 37, 75, 76, 155, 162
Pound, Roscoe, 133

Q

Quork Walker Case, The, 41

R

Rationalia, 63, 64, 65, 66, 81, 83, 121, 153, 157
Reign of Terror, The, 37, 155
Replication Crisis, The, 95
Robespierre, Maximilien, 37, 38, 39, 40, 41, 43, 47, 50, 124, 155, 162
Rogues Island, 25
Romanticism, 33, 34, 35, 42, 43, 164
Rousseau, Jean Jacques, 35, 37, 38, 42, 50, 162, 164
Rylands Library Papyrus P52, 104, 105, 106

S

Scalia, Antonin, 21, 22, 58, 148, 149, 156, 163
Separation of Powers, 30
Smith, Adam, 20
Socrates, 2, 6, 16
Solzhenitsyn, Alexander, 132
Soviet Union, 7, 11, 21, 30
Spooky Action at a Distance, 114
Statistical Fallacies, 107

T

Taft, William Howard, 53
Tennis Court Oath, The, 35
The American Scholar by Emerson, 42
The Humanitarian Theory of Punishment, 131, 135
The Milgram Experiment, 139
The Model of the Jury System, 147
The Moral Landscape: How Science Can Determine Human Values, 65

The Population Bomb: Population Control or Race to Oblivion?, 115
The Postmodern Condition: A Report on Knowledge, 77
The Problem of Faction, 23, 27
The Quest for the Historical Jesus, 102
The Study of Administration, 47
Tyson, Neil deGrasse, 63, 64, 65, 66, 68, 74, 81, 83, 153, 157

W

Wakefield Vaccination Study, 93
Wanderer above the Sea of Fog, 33
Washington, George, 28, 30
White Logic, White Methods: Racism and Methodology, 80
Wilson, Woodrow, 8, 9, 46, 47, 48, 49, 50, 51, 52, 54, 56, 59, 60, 61, 123, 124, 126, 133, 150, 164

Λ

λόγος, 75, 78

www.ingramcontent.com/pod-product-compliance
Lightning Source LLC
Chambersburg PA
CBHW061839300426
44115CB00013B/2447